Drilling for Meteorites

By James P. Tobin

© 2017 James P. Tobin

ISBN-13: 978-1979252898

ISBN-10: 1979252890

This is an original work, all rights are reserved. No portion may be reproduced by any means mechanical or electronic without the express written permission of the author. Except for limited quotations for legitimate review and educational purposes.

Other than the Elihu Thomson portrait, the four images taken by L.F.S. Holland and the two pages from the Fairbanks Moore Catalogue, all other images are original scans or creative work by the author with the one exception of the image on page 179 of the tunnel opening and timbers. It was generously supplied by Paul Harris.

To Sara
My friend, my love, the joy of my life

Table of Contents

Introduction..7
Forward...11

The Times
Meteor Crater in 1920 America...13
Daniel Moreau Barringer's Motives..22

Barringer Meteorite Crater
The Crater..25
The Crater Floor..32
The Meteorites...39

The People
L.F.S. Holland..53
D.M. Barringer...58
William Francis Magie...64
Elihu Thomson...67
Sidney J. Jennings...71
The Foreman Mr. Wammock..76
The Visitors..81
The Drilling Crew...85
The Company...92

The Work
Preparing the Drill Site..97
The Water and the Pipeline..107
The Rig..115
The Magnetic Survey...126
The Drilling..142
The Drill Log..163
The Tunnel...171
Crater, Arizona Post Office..180
The Money...184
Barringer's Footprints..192
L.F.S. Holland's Own Words..208

The Aftermath
What Remains..226
L.F.S. Holland After Meteor Crater..231

* * *

Appendix One...236
Appendix Two..240

Introduction

There are moments in life when you get to fully realize the ultimate pleasure an intellectual pursuit or hobby has to offer. To find meteorites where they fell and then hold them an instant later in your hand. Those are such defining moments. Or to go to a location no one has studied and be the first to find evidence of ancient occupation. I have experienced that as well. When first presented with the files of L.F.S. Holland I had another of those completely fulfilling experiences.

Books come from many places. Some come from personal experience, some from powerful imaginations. Most require at least a little research. A few books are the story that flows from the research materials themselves. It was February 2014 and Paul Harris and I went to the Tucson Gem and Mineral Show. Every year finds us there for a few days to meet and greet and buy meteorites. While in the showroom of friends the conversation turned to Meteor Crater. There was mention of a collection of old papers from the 1920's about work done at the crater. I expressed my deep interest in at least being able to see the documents. The following year the conversation repeated but with the real possibility that a meeting could be set up for me to borrow the documents and copy them. Scheduling problems and even bad weather caused all those plans to fail, and it was arranged that I could get the papers at the gem show in 2016.

I did not know how many pages were in the collection of documents but knew that their safety was the most important thing I had to worry about. We visited Pieter Heydelaar and Debra Morrissette's room several times during our days at the gem show. We told them that we would take the papers the last thing before we headed to the airport and I would carry them with me. I put all my clothes into a box of meteorites we were shipping home. That made room for the files in my carryon bag.

Once home I began to go through Mr. Holland's records, and each page I turned was like another Christmas present to open. Holland's year of work which has never been reported with more than a few lines was told in rich detail. There was a piece of history that needed preserving. I scanned the documents and printed out copies. I interleaved the century-old paper with buffered archival sheets to protect them. I burned a disk of the scans and put all the originals into folders made of acid-free paper. Two ream size boxes held the copies I had printed out. My wife and I made a road trip of returning the originals and copies to Pieter and Debra. Just a couple weeks after getting the precious documents they were on their way home.

In one sense I realized from the start that it was not the complete larger story of the whole drilling program on the south rim of Meteor Crater. Mr. Holland was there for only a year, and the entire enterprise lasted well over two years. But I knew many smaller stories were complete, embedded there within the greater tale of disaster and struggle. Since these were the personal files of one man, they were complete as far as his work was concerned.

Over the next month, I just read the documents and tried to figure out a scheme to organize the materials. After I had a little handle on the scope of the story, I began to put down words. I quickly discovered that I was spending too much time searching the piles of documents to find facts. I can usually get by with just some notes scribbled on a sheet or two of paper for magazine articles. But for this project there were so many details, facts, and dates about which I wanted to write, I was forced to use index cards.

The world has changed considerably since 1920, and nothing at first glance would seem to be the same. But, it quickly became clear that men are men and little which drives them changes much with time. The motivations, as well as the attitudes of the characters, were much like those nearly a century later. People remain a mixture of good and bad, of genuineness and prejudice, sometimes demonstrating patience and other times plagued by anxiety.

There is an unedited quality to the documents. It is easy to believe Mr. Holland put them away and never read them again. Some of the content, reflects the negative side of man. Speech and manners common in 1920 are at best archaic today or at their worst offensive. We hope as individuals that we will live transparent lives. That who people see in public is who we truly are. Some of the documents that are not related directly to his work have a different tone. Letters written to acquaintances are rougher in content from those written to bosses. The topics are on a few occasions barely fit for re-discussion. While I mostly admire the man for his honesty, perseverance, and hard work, his aristocratic views, and disdain for certain people are not something I admire. But it is part of this story that Mr. Holland held such views. That these occasional bits of the darker side of the man still exist in the collection gives the whole collection some added authenticity. He never removed anything later that he thought embarrassing or unseemly. I can visualize him packing to leave Meteor Crater on May 1, 1921, grabbing the files and placing them in his luggage. Upon arriving at home putting them in a filing cabinet and never thinking of them except rarely during the next thirty-odd years.

We look at the Moon, we see the light and dark areas forming the Man in the Moon, the character of story and legend since ancient time. But, we see no craters. Gaze at the Moon through even a small telescope and countless craters will fill the view. There is no atmosphere on the Moon, no weather, so no erosion. Every asteroid and smaller meteoroid that strikes our neighbor forms a crater. Every impact happens at full cosmic velocity. Until a crater is finally covered by the debris of many others forming nearby, it will remain. On Earth, small stones that reach the ground will penetrate just inches or stay on the surface. Any small impact pit created by a stone of a few ounces or pounds will disappear with the first rain that follows the fall. Larger asteroids form craters that last for geologically long periods of time, but finally, they erode away also. Only the evidence of damaged rocks far below the impact site may still survive.

As of this writing, there were 190 officially recognized impact craters on Earth. Most of the thousands of craters from the past are gone completely. Some number with faint remnants still await discovery. Most astronomers a hundred years ago turned their telescopes upon the Moon and believed that the craters seen there formed by volcanic activity. Today the lunar craters are known to be from impacts of asteroids and meteoroids. Arizona's Meteor Crater played a significant role in the changing of this view. There are volcanos on other bodies in our solar system, but we know that most of the craters are the result of collisions.

When L. F. S. Holland went to Meteor Crater in 1920 the debate over its origin was still going on. It had been an argument within the community of geologists and astronomers for already two decades. But by the end of the 1920's it drew in mathematicians, physicists, engineers, more astronomers and more geologists, becoming a topic of debate in magazines whose readership was the general public.

In the early years of the debate over Meteor Crater's origin, there was no information to use for a starting place. When a sinkhole was discovered in the past the reasons for its formation were understood. Sinkholes had been seen around the world for hundreds of years and studied. The same was true for volcanos, hundreds were known even in the 1890s, and many had been investigated. But no person had ever studied an asteroid impact crater until Meteor Crater was investigated.

The drilling program supervised by Mr. Holland was just one of several attempts to find the buried asteroid. It was a financial disaster lasting over two years. A hundred years later the effort is summarized in most histories of Meteor Crater in a single paragraph with none of the participant's names mentioned. Even in scientific works about Meteor Crater, the drill log of the second supervisor who replaced Mr. Holland is often the only document used as a resource. Never finding the asteroid may explain why nothing from the activity is used by scientists. But there was a human story important to the history of the crater. A drama about men struggling to work at an isolated mining camp in 1920. That story is contained in the papers of L. F. S. Holland. Historical nonfiction is not every reader's cup of tea. But, sometimes the story is dramatic, and the scenes are filled with famous people. Then historical books can serve to shine light upon a time in the past when life was very different.

A note about the writing.
Much of the content is in the form of directly cited quotes from the letters, telegrams and weekly reports of L. F. S. Holland and others. The quotes are reprinted without correcting the spelling, punctuation or other errors they contain. It seemed appropriate to present the world as it was, and the writers as they were, persons who spoke differently and made mistakes. They were individuals that used archaic language and styles. It is confusing at spots where words such as "instant" are written for no apparent purpose. But that was the method in 1920 to tell the reader that they have just received

a communication and are replying immediately or to state that some event is occurring now. The manuscript has been edited, and great care has been taken to eliminate errors by this author. But the reader will see errors in the quotes of the characters of the story.

A message from the document owners Pieter and Debra Heydelaar

A few years ago we acquired a collection of documents from the personal collection of Mr. L.F.S. Holland superintendent of the Barringer Crater operations during the years 1920-1921.

The collection consisted of an incredible amount of information-including several letters from Mr. Barringer himself, purchase orders for equipment, photographs, receipts, etc., material never before published.

The story was there of the day to day operation with all its problems, breakdowns, frustrations, like so many mining operations. We knew we had something special and wanted it preserved and to be able to share it.

We had known Jim Tobin for a long time, like us a meteorite enthusiast with a great interest in the Barringer story; he is the author of the book "Meteor Crater" so with his knowledge and enthusiasm about the history of the Canyon Diablo Crater what better person to ask to facilitate the preservation process. He readily agreed, copied and scanned the original papers then wrote this fascinating book based on this new information.

Thank you, Jim, for all your work, done with much enthusiasm, through your book this great story will live on.

Pieter and Debra Heydelaar

Foreward

It was going to be a hot sticky day, and I was going to be digging and raking rocks if the Saturday followed the normal plan. My parents were building a house in Bullhead City, Arizona and I was about 15 years old. This was not my first trip from LA to the Bullhead house. We went out to work nearly every weekend. We had done so for a few months. The pattern was always the same. Pick up my father at work Friday afternoon then drive to "the river" as my parents phrased it, and work all day Saturday and half of Sunday, then drive home. My task for the last few weekends had been digging the great hole for the septic tank and spreading the rocks I threw out of the hole over the rest of the property. And there were a great many rocks to throw and rack. The house was just two short blocks from the Colorado River on an alluvial fan made of mostly cobbles.

I had not gotten used to sleeping in the humid air of the swamp cooler; it was really still too hot for me as a kid from near the ocean. My mom had breakfast on when I got up, and I had mentally come to terms with what a rough day it was going to be. In very uncharacteristic style for my parents, they said to get dressed in something nicer after finishing breakfast for they had a surprise for me. Soon we were all back in the 1960 Ford F-150 pickup truck with the white camper heading to Meteor Crater for the day. I was in shock. I had by this time already sent samples of some odd rocks I had found to Dr. H. H. Nininger to see if they were meteorites. I had been observing the Perseids each August sending counts and charts to Sky and Telescope. But getting to go to Meteor Crater was the best.

I hiked the circumference of the crater that trip and not down into the interior. I had my choice, and going around, was it. I was so excited I could have screamed. Everything I had been reading about was in front of me to touch and see. I was hooked. I would return to Meteor Crater many times after that as an adult. When I wrote my first book on the general history of the crater, I stayed there for a week in the RV park. I went to the crater twice a day. They marked my first admission so I could come back free for the rest of the week. I took about ten rolls of film and drove into Flagstaff to have them developed before returning home. I needed to make sure the pictures were good. It was just a little book, actually only a folded saddle-stitched booklet. But, I sold hundreds of them at star parties and astronomy gatherings. And writing it fanned the flames to learn more about the crater. Two bigger books on the crater since and my passion for learning more is undiminished.

With that background, it should be easy to image how thrilled I was to have a chance to pour myself into the day to day records of the 1920's drilling program. I was out hunting meteorites at Holbrook, Arizona while writing this book and of course, I stopped at the crater on that trip. I can not go down I-40 without stopping at Meteor Crater. It is not that I see anything new when I visit or take a better image with my camera, it is more that I still have the same passion I had when I was a teenager for anything and everything meteorite.

The Times

Meteor Crater in 1920 America

It has been three years since Moon Pies were invented and two years since the end of World War I. There is a boom going on in the United States as often happens after winning a big war. It is the beginning of the Roaring Twenties for us, and it is the "années folles" or Crazy Years in France. Times are not so good in Germany as its people suffer after the war. America is not joining the League of Nations.

The 18th amendment prohibits alcohol production, and the 19th amendment gives the vote to women. The Russian Civil War is over, but the famine begins, and communism spreads. Silent films will almost end before the decade does and color motion pictures will appear for the first time.

Commercial radio stations have been operating for about a month when United States Smelting Refining and Mining Company sends Mr. Holland out to Meteor Crater. But, he will hear no broadcasts or even have a radio. Both of the radio stations are back east, and he is in Arizona. Which in many ways is like being on another world or in another time.

Before World War I active mines operated all over the western states. Miners were in good supply. You can see the remains of the mines everywhere in deserts and desolate mountains from Colorado to California. The men that had worked them went to war, and when they came back, they did not return to hard rock mining. They found jobs in the cities of the east. For those needing labor in Arizona in 1920, it was tough to find men that wanted to work.

Getting labor for the site preparation was among the chief problems for Superintendent Holland. Not far behind was getting supplies. Much of it came from either the east coast or from Los Angeles or San Francisco. It all came by a freight train. Even some of the food for the workers came from a wholesale grocer in Denver. Everything had to be unloaded from the train and trucked to the crater. And it had to be moved promptly, or they would incur charges for demurrage. Whatever you shipped had to be retrieved from the train yard at the destination. If you delayed a freight car by not unloading it, the railroad charged you. The Crater Mining Company had an automobile and a Ford truck. They also used a local transport company on occasion. Holland makes a statement in his weekly report to the home office dated October 23, 1920 that he was "glad to say that so far we have not had to pay a cent for demurrage on cars of any material delivered."

Adults tell their children tales of what gasoline cost when they began to drive. Filling a car with gas for the first time was like a rite of passage for American youth in the 1960s. We reminisce about the good old days when things were cheaper. But in truth, it was hard for people to pay for what they bought in 1920. While on the subject of gasoline, Superintendent Holland is going to need much more gasoline once the drill rig is complete because he has been given a gasoline engine to power it.

Up until the site preparation work was finished, he only needed fuel for the car and truck, but soon he will be running the rig engine 24 hours a day. Gasoline has been costing 34 cents per gallon F.O.B. Holbrook, Arizona. The price increased as of October 1920 to 36 cents a gallon. The fuel comes in barrels from Standard Oil. It costs an additional 2.54 cents per gallon to have it shipped from Holbrook, Arizona to Sunshine Station near the crater. If he orders a tank car load of gasoline from Standard Oil, the price is 35.2 cents per gallon. The volume discount savings is a meager eight-tenths of a cent per gallon. They would also save the 2.54 cents charged for forwarding the gasoline from Holbrook. However, the tank car shipment would take three weeks to arrive directly from Texas and Oklahoma refineries. Crater Mining Company would have to buy a trailer with a special tank and a larger truck than their Ford to haul the fuel from the train station. They would also need storage at Sunshine Station as well as at the crater. Holland wisely recommends to his employer in Boston that they continue to get the gasoline in barrels which their current truck can handle and pay the 38.54 cents per gallon.

The lumber for the drill rig came into San Pedro Harbor south of Los Angeles and was freighted on to Sunshine Station. Republic Supply Company manufactured the drill rig and derrick. Two rigs were purchased. They believed at the beginning that they would be drilling more than one hole. The contract was for up to ten holes. When assembling the first, the riggers found that there were missing materials. Wood from the second rig was taken and used on the first rig as were many of the bolts. The materials had been inspected and a supplementary order placed, yet they were still far short of all the parts when the erecting took place. The cost for the two rigs was just short of $30,000, and they could complete only one. It made little difference in the end. There was just the one drill site with all of its problems. The riggers told Holland that it was their common experience to find insufficient lumber sent to everyone except the biggest companies like Standard Oil. The big operators did not receive orders with shortages. It seems that bad business practices, cutting corners and treating smaller customers poorly are not new problems. For example, Mr. Holland had the riggers build the bullwheel and the pitman of the second rig so they could be available for use on the first rig should those parts break. He was amazed to find shafts made of pine wood, not iron or steel. The pitman was also of pine instead of hardwood. He expected that the rig was strong enough for anything required in their churn drilling. In this he was nearly correct rarely did problems with the rig itself arise.

No evidence remains at the drill site today of the deep pit that was blasted out under the rig location. This cellar was for handling the long lengths of casing and drill rod. It was intended to be blasted out with dynamite but instead was blasted with black powder. The dynamite never arrived after weeks of waiting. Which based on the condition of the roads to Winslow might be a good thing. All three of the roads were appalling and needed work constantly. Holland makes mention in a weekly report of the repairs done to the Ford truck after it broke down on the rough roads. Whenever possible, they wanted materials to come to Sunshine Station. It was much closer, and they had what seems to be a good relationship with the railroad agent. They had an acceptable relationship with the manager at Volz Trading Post too though it was farther away down at Canyon Diablo Station. Great was their desire not to travel any farther on the bad roads than necessary.

Holland and Barringer petitioned the Postmaster General to establish a post office nearer the crater so that mail could be received more frequently than once or twice a week. They were paying for a driver to bring the mail out along with newspapers and telegrams from Winslow. Eventually, a closer post office named Meteor, Arizona opened at Sunshine Station. The postmaster in Winslow agreed to put the mail on the train in a bag that would be taken off at Sunshine Station.

Below is the present day railroad crossing at Sunshine Station. Just a few yards away can be seen the remains of the building foundation and the old loading dock that was there a hundred years ago. The following image is a small amount of the debris which remains from that time. The rolled and riveted pipe that still lies around the station site is of the same type that was originally used at Meteor Crater. By the time of Mr. Holland's arrival, the pipe in and around the crater was mostly so rusted as to be unusable. He was forced nonetheless to salvage what he could from both the crater floor and the existing old pipelines on the surrounding plain.

Little evidence shows in satellite imagery of Sunshine Station except for a suspicious white spot of ground that is seen in the image above. The faint remains of a building location can also be seen in images taken from space. A careful observer visiting the site today will find in addition to pipe, bits of broken glass and crockery, wire and half buried cables of steel rope. But, there are none of the ruins of buildings, so commonly seen at other sites near the crater. Holland made shipments of the stark white silica sand during his time at Meteor Crater. Barringer was always trying to sell any of the crater's resources. For several years in the late 1940s to the early 1950s, the property of the south rim was leased to mine the silica. Meteor Silica Corporation sent hundreds of tons of the silica sand to The Ball Brothers Glass Company in Los Angeles. These shipments from Sunshine Station may explain the very visible white spot that remains there today.

The gasoline engine for the drill rig was a constant source of problems for the entire first month of drilling. Crater Mining Co. was communicating with the Union Tool Co. of Torrance, California about the difficulties. There were no documents to indicate for certain that the Ideal Engine was bought from Union Tool Co. but for sure they were the ones who did service and repair on it. At one time they were holding a second gasoline engine for Crater Mining Company but found another buyer and let it go. This might indicate that they had indeed sold the first engine. Union Tool Co. would receive a poor score on a modern customer survey from Superintendent Holland. In December of 1920 Union Tool Company was bought by National Supply Company and Holland hoped that service would improve. He wrote "it can hardly be worse than we have had hitherto. For example, they have sent no less than four defective fuel pump bodies for their gasoline engine, and we had to rebore the final one ourselves to make it fit the engine." They ultimately sent a man after many complaints, and he got the engine running slightly better. Further tinkering and repairing by personnel at the crater finally got it consuming the proper amount of gasoline. But, supplies, maintenance, and labor were not the only problems the crater's location presented.

On Monday night September 13, 1920, Mort Branson who was to be the drilling foreman was stricken in his bed with paralysis. He was in critical condition when taken over the terrible road to Winslow. The doctor there advised Superintendent Holland to send him to Los Angeles. On Tuesday morning he was put on the train accompanied by Holland's best worker. Branson's wife met the train on Wednesday in Los Angeles. We learn later that his condition had much improved. However, his illness delayed the work further. Branson was personally chosen by a Mr. Coan of a California-based oil company. Furthermore, Branson had arranged to hire his drilling team from good men with whom he was familiar. Now another foreman had to be recommended by Mr. Coan and another team of drillers chosen.

Meteor Crater was quite literally out in the middle of nowhere in 1920. The short 5 mile trip off I-40 today is immensely different from the isolation people experienced at the crater decades ago. Mr. Branson was not the only person to need medical care.

Workers who got ill would lay off work in the bunkhouse until they got better, those more seriously ill would quit or be fired. Most places with no doctor nearby relied upon someone who served as a first aid giver. The mucker who smashed his toe was undoubtedly treated at the crater and either later sent to the doctor in Winslow or as we say today had to suck it up and go on with life hoping it healed well enough on its own. This writer's father chopped the great toe off one foot with a post hole digging tool during the same period of American history, except in Oklahoma. There was no doctor there either. His toe though reattached by someone in his hometown was never right afterward, but it healed. Many similar stories exist about 1920s rural medicine. During the digging of the tunnel into the south wall of the crater, the men suffered serious problems from breathing the silica dust. Holland provided respirators, but the miners often did not use them, considering them unmanly. Several men quit or were laid off because of bad lungs. Workers had little protection on the job in 1920 other than the hope their employers were moral and ethical individuals. Barringer had experienced the same problems with the dust when the first shafts and trenches were dug. Some of his men began spitting up blood from the silica dust. The impact shattered the sandstone into sharp angular particles. He was very concerned that Holland was aware of the dangers of breathing the dust. Barringer made an inquiry to the Bureau of Mines about respirators. The reply letter from that branch of the Department of the Interior is among Holland's papers. But Holland had already gotten respirators from Goodyear Tire and Rubber and would later add a blower, air tube and watering system to clear the air and hold down the dust in the tunnel.

On one occasion the County Sheriff was required to come to the crater to deal with a dangerous situation. There being no telephone at the crater meant calling him from Winslow, Canyon Diablo, or Sunshine. Someone had to drive there and make a call or send a message. A laborer Holland hired in a batch of a dozen men was insane and caused trouble with a rifle. He had to be removed by the sheriff. It is not hard to imagine a Hollywood western movie scene. A struggle to disarm him followed by tying him up and waiting for the lawman to come. Arizona was still in many ways the wild west in 1920.

Statistics and facts never make for exciting reading, but it seems necessary to briefly describe the way Arizona was at the time that Mr. Holland went to Meteor Crater. The area of Arizona that Meteor Crater occupies is in the southern portion of the Colorado Plateau, a vast area covering parts of four states. The elevation of the plateau around Meteor Crater is 5500 feet above sea level. Approximately 12,000 years ago humans found the area and settled along the Little Colorado River. They were hunter-gather bands that would in time become the founders of the Anasazi Culture. They would settle across much of the Colorado Plateau. For reasons still not fully understood the Anasazi left the region and their established villages. Like the Maya, these people did not truly disappear, they moved and abandoned their old ways. The Pueblo peoples living in the southwest today are descendants of the Anasazi. Had they used up the land and over populated the area? Was it a nearly thirty year period of drought from 1270-

1300 AD that stressed the culture to the breaking point?

There were several attempts by pioneers from the east to create settlements in the area of Meteor Crater along the Little Colorado River. Most failed after a few years. Winslow, Arizona grew out of the one surviving early settlement because of the railroad. Named for Edward F. Winslow the President of the Santa Fe Railway it was thriving by 1920. Flagstaff, Winslow, and Holbrook were all towns with only a few hundred persons by the 1880's, but the railroad changed all that becoming the big employer and the provider of opportunity.

Coconino County when first established in 1891 had a population of just 4,000. The Chamber of Commerce in Flagstaff formed the same year. One of its first actions being the promoting of an excursion from Chicago to the Grand Canyon. 1891 was also the year that Grove Karl Gilbert made his investigation of Meteor Crater. Gilbert would travel to the crater from Flagstaff where he had arranged for wagons, horses, and drivers. It would take him three days to reach Meteor Crater just 35 miles away. There was a train station at Canyon Diablo much closer but no wagons, workers or sufficient supplies apparently. He had to take all the food, water and equipment for the study of the crater which would last seventeen days and he had a staff of men with him. He was also going to stay in the area of Flagstaff after the Meteor Crater survey to examine the volcanic region near the city.

The Santa Fe, Prescott and Phoenix Railroad was completed in 1895, and the area began growing at a faster rate. Even as early as 1886 Flagstaff was the largest town between Albuquerque, New Mexico and the Pacific Coast. But the population of Arizona was still small even by the time Holland arrived in 1920. Flagstaff would report only a population of 3,186 that census year. Holland would have constant problems finding workers in the early months of site preparation because of the low population. The entire Coconino County had a population of just 9,982 in 1920. Second, in size behind San Bernardino, County in California which is the largest county in the nation the workforce in Coconino County was spread very thinly.

Winslow, the closest real town to Meteor Crater had a population of 3,730 in 1920 and was a major railroad stop. Service yards, train yards, and warehouses were built parallel to the tracks with the city growing up around and away from the tracks. Holland would hate the drive into Winslow because the roads were terrible most of the time. But in truth, it was the closest place for any support he and the workers might need. It was in Winslow that the casual laborers spent the few dollars they received for their work at Meteor Crater. It was Winslow that hosted the celebrations that the local cowboys attended who rode for Mr. Hart, the cattleman. He worked the land around the crater and suffered from the same labor shortages as Crater Mining Company. At one point during 1920, Hart had just a single cowboy in his employ.

Holbrook, Arizona 60 miles from the crater had a population of just 1206 in 1920. In

1910 the population had been only 609. The most exciting thing to happen in Holbrook between those years was the famous meteorite fall of July 19, 1912. Tens of thousands of stones fell at the Aztec rail yard east of town. It remains one of the largest meteorite falls in history. A couple of dozen people from that tiny population of Holbrook traveled out from town to hunt for the stones. They immediately recovered over 14,000 meteorites most of which were sold to museums and mineral companies in the east.

The entire state of Arizona's population in 1920 was 334,162. That was about one-tenth the population of California at the time and about four times the population of Nevada. Phoenix had the largest number of persons in 1920 with 29,053.

The climate of the area was and remains diverse. In the winter there could be snow to a great thickness in Flagstaff, and there could be rain across the region. In the months of January, February and March of 1920 there had been significant rain, as much as eight-tenths of an inch in a single day. But after Mr. Holland's arrival in May, there was just one more rain caught by the dams across Canyon Diablo. The remainder of 1920 was extremely dry. Flagstaff has never recorded a temperature of 100° F. It is in the mountains. But, just 35 miles away out on the plains Meteor Crater can often present dangerously high temperatures to the unprepared person.

By the mid-1920's the "old highway" would be completed which connected the small cities of northern Arizona. It would later become part of Route 66. Still, later I-40 would replace Route 66 during transcontinental highways construction of the 1950's. Some of the small communities and businesses that had been along Route 66 would

close their doors and become the ruins that can be seen from I-40 today.

Very near Meteor Crater is a unique ruin that once housed the meteorite collection of Harvey Harlow Nininger. Its tower has collapsed, and the walls made of local Moencopi sandstone have fallen partially down. Travelers along Route 66 would stop and tour the museum for a twenty-five cent admission. Meteorites and books were available for sale. Near the end of his time in the building, Nininger had the largest private collection of meteorites in the world. He has often been called the father of meteoritics in America. After his museum had been bypassed by the creation of I-40, so few visitors came that he was forced to move to Sedona, Arizona and reopen his museum and store there.

Canyon Diablo which was both a train stop and a tiny community is now just a scattering of old foundations and building sites. They show well on satellite images, but no structures remain. The same goes for Sunshine Station, the preferred railroad stop for Mr. Holland and Crater Mining Company. It was just about 6 miles from the crater. Most of what Crater Mining Company received came there. Nothing remains at Sunshine either except the normal bits of refuse that are always left when old occupation sites are abandoned.

1920 was a tough time to try and do anything at Meteor Crater. It was isolated, and there was only the railroad to provide the essentials of life and work. There were no luxuries and few conveniences. No telephones were at the crater. Communication was done by the US mail and by Western Union telegrams. The men probably played cards and amused themselves with what activities they could create. They had only a little time off. It was common to work seven days a week and to work 55 to 60 hours each week. Back-breaking physical labor of nine decades ago has been much replaced by machines. Things are much safer today. Workers are protected by more than the hope that their employer is a thoughtful and concerned person. There were few laws in place to regulate working conditions in 1920 though such laws were soon to come.

There was the dark night sky for the men at Meteor Crater to enjoy. To sit under the canopy of the stars in the pitch black darkness and see the myriad of tiny lights. Meteor Crater's great chasm next to them in the blackness. Were they too tired to enjoy such a thing? The drillers knew why they were there. They knew the goal was to find a fallen star buried below the south rim. Did they fully understand the significance of being the first to make such an attempt? Mr. Holland knew, and D. M. Barringer knew that they were doing something never done before. It can only be guessed what the drillers truly thought. Did they think their employers were crazy or foolish? We will never know for sure, and they are all long gone as is most of the dark sky in America. But Meteor Crater remains isolated even today, and the stars still shine, and the Milky Way is still visible especially on moonless nights. Meteor Crater still reminds those gazing upward into the darkness that sometimes objects fall to Earth with terrible consequences.

Daniel Moreau Barringer's Motives

Daniel Moreau Barringer has gone down in history as a pioneer of impact theory and the original student of impact craters. Meteor Crater was the first of the Earth's craters to be investigated and what Barringer learned and wrote was quickly used to study other craters such as the Henbury craters in Australia and the Odessa Craters in Texas. Some of his errors and misconceptions traveled to those other craters as well. For example, a vain search for the buried mass of iron was conducted at both the Henbury crater site and the main Odessa crater. It took a long time to accept the vastness of the energy asteroid impacts release. So it took the same amount of time to understand there was not going to be a large surviving mass of asteroid. There was a lengthy debate about Meteor Crater. The volcanic steam explosion theory took a long time to die out. In the end, Barringer was proven to be correct.

As much as these accomplishments are outstanding and well deserved, the focus of his thoughts in 1920 was not on discovery or scientific knowledge. His mind was on the exploitation of the crater for profit. Meteor Crater was nothing more than a ready-made open pit mine. If there had been vast reserves of surviving meteoric iron under the crater floor, he would have dug it out. He would have done so with little thought for the destruction of the crater's geological uniqueness. 1920 was not a time in which private industry felt restrained by environmental, historical or archeological considerations. If any of the money making plans and schemes for the products of the crater had come to fruition, then there would be no Meteor Crater as it is today.

Daniel Moreau Barringer died November 30, 1929. The family members had pledged never to sell the property and to continue to explore for the buried asteroid. The family clearly understood it was his wish that once it was located and could be mined profitably, his descendants were to do just that.

When efforts to find the buried iron would stall, Barringer would continuously try to sell the silica sand. Repeated attempts during his lifetime never materialized. Frequently samples of the sand both great and small were shipped off to prospective users. Finally long after his death, a deal was struck. For a short period beginning in the autumn of 1946 a lease for mining the south rim was arranged. Meteor Silica Company formed. For about three years the company worked at selling the silica. Most was sold to The Ball Brothers Glass Company of El Monte, California. Hundreds of tons were removed from the south rim where it is exposed in an arroyo and easy to mine. Shipments to The Ball Brothers Glass Company stopped upon the discovery that the fineness of the silica flour allowed it to blow up the stack of the plant. It was contributing to the poor air quality of the Los Angeles basin. In the early 1950's The Barringer Crater Company had the rights again to the property. Feeling the need to pay some attention to the criticism about their defacing the natural beauty of the crater The Barringer Crater Company stopped offering any leases for mining the silica.

To depict Daniel Moreau Barringer as a man wholly motivated by economic gain in his activities at the crater may seem harsh but for the most part, it is the truth. Nearly all of his interactions with scientists revolved around establishing the amount of iron buried at the crater, and where it lay. Barringer would use the information the scientists calculated to promote the crater as a rich iron resource and to secure investors for mining the metal. The personal returns to Barringer of notoriety and recognition as the discoverer of its asteroid impact origin were without doubt bonuses. But, his fundamental motivation was to get the iron out in a commercially profitable manner and amass wealth in the process.

At that time in American history, there was nothing unusual about having such goals. The world was here for man's use and exploitation. It was the wise man that could discover mineral resources, stake a claim and mine them. It is fortunate that the iron was not there and that the other products were so difficult to recover or were often commercially unsuitable. Otherwise, the crater would not be today an excellent laboratory for impact crater research.

In the course of all the activities done at the crater, dozens of shafts and trenches were dug. Dozens of holes were drilled in and around the crater. The dump site of stark white silica in the crater center is the most visible reminder of the work within the crater. The silica mine on the south slope is the most visible scar of the work on the exterior of the crater. But in fact, there was a massive amount of material moved for the south rim drill site. Approximately five thousand tons of the ejecta and red sandstone were removed to prepare the 100-foot by 40-foot drill site. The red sandstone was drilled and blasted to a depth of 25 feet at one end of the area to level the site. Additionally, a cellar was blasted 18 feet deep where the drill rig would stand for handling the long lengths of casing and drill rod.

The very visible streak that runs down the face of the south cliff like a rusty waterfall is the result of the drilling on the south rim. It was originally a brilliant white streak that was truly ugly. After nearly a hundred years of rusting and washing away, it looks much more natural. Today it is a little difficult to find the drill site location on satellite images unless you know where to look. Fortunately, there are plenty of clues to help find it. The horizontal tunnel tailing pile, the silica mine, the remains of the old roads up on the slope can aid in locating the spot the drill derrick occupied. There have been nearly a hundred years of water erosion and wind, the contours of the leveled area have softened slightly. Stripped of the gray colored ejecta the red sandstone of the drilling site remains visible. It marks where a great struggle of man, crater, and machines took place. Barringer may not have been in charge of the operation in 1920, but the men who were in charge shared his financial motivation and cared little for the priceless scientific value of Meteor Crater.

The drilling work on the top of the south rim in 1920-1922 was one of the most expensive explorations for the buried asteroid. It would lead to an even more costly attempt to reach the phantom deposit of nickel-iron in 1928. Barringer did not live long enough to see the final failure of the last shaft dug on the south slope. But, that attempt was based on what he believed had been found in the drilling program on the edge of the south rim those few years earlier. Today gazing backward in time to the evidence the drill brought up there just was not much to warrant the far greater expense of the final shaft. But, Barringer was a man clearly determined to find riches buried at Meteor Crater. His mind was shut to the realities being learned even then about asteroid impacts. However, there is no doubt he was a persuasive and relentless salesman. Given some time he always found people looking to get rich with him.

This is a small scale pan with silica rock flour from the south slope of Meteor Crater.

Barringer Meteorite Crater

The Crater

Nothing quite describes the experience of coming around the corner of the museum building at Meteor Crater and seeing the immense cavity for the very first time. If you are a little acrophobic, it can create that anxious feeling and tightness in the stomach, especially if you get too near the edge. Meteor Crater is not a large crater. Compared to some in our solar system it is a tiny crater. We see the thousands of craters on the Moon easily with even a small telescope. But each of those is larger than Meteor Crater which would not be visible if one attempted to view Earth from the Moon with that same telescope. Standing on its rim and gazing across the 3,900 feet to the other side speaks volumes about the power of natural events.

On some otherwise normal day 50,000 years ago an asteroid entered the Earth's atmosphere. In a few seconds, it touched the Earth's surface at what would later be called North Central Arizona. The asteroid continued into the ground until the resistance of the rock was finally able to bring it to a stop. At that moment the energy of the asteroid's cosmic speed and the compression of the rocks that were squeezed and crushed as it penetrated the ground was released. The hundreds of feet of rock above the focus of the explosion were shattered, thrown upward or flipped over onto the land surrounding the rim and pit which formed almost instantly. A fireball and ground wave moved outward from the point of impact. Everything living in the area for many miles was killed and burned up. The ground wave was felt for a much longer distance.

Of course, no humans were living in the area when the crater formed. And we do not know when the crater was first seen by humans. From narratives of the last century we know that the Native Americans avoided the crater. They also wanted nothing to do with the meteorites found in the area of the crater. In the 1940s Harvey Harlow Nininger had his Meteorite Museum on Route 66 in view of Meteor Crater. Elders of the local Native American tribe would prevent younger members from entering his museum telling them that their people have nothing to do with the meteorites or the place they fell. At some point long in the past man did find the crater. It is interesting to wonder what they might have thought. When a hiker climbs the slope of Meteor Crater, there is no clue that a great chasm is just ahead. It is only when you finally reach the top of the rim that the expanse of the crater is revealed. From a distance, the rim of the crater simply appears as a low ridge.

By 1891 Meteor Crater had been discovered and was called by many other names. It had also caught the attention of the geological community. Grove Karl Gilbert the head of the US Geological Survey took his trip to Arizona to investigate the crater. He had sent Willard Johnson out months earlier to study the site. Johnson's report that it was formed by a volcanic steam explosion seems to have only heightened Gilbert's own desire to see the crater. Gilbert began with the belief that it could have been formed by an asteroid striking the ground. But, when he found no magnetic evidence of buried iron and could not reconcile the amount of rock around the hole with the volume of the hole itself he also decided it formed by volcanic processes. This decision and his article about the location would lead to a great debate lasting for decades to follow. Gilbert would never involve himself in that debate and would remain silent about the crater for the remainder of his life.

In the years immediately after Gilbert's visit, others would come to believe in the asteroid impact origin of Meteor Crater. G. P. Merrill, a geologist from the National Laboratory, would write a paper declaring the impact origin. He would describe the crater and its meteorites in great detail getting some of his information from communications with D. M. Barringer.

Daniel Moreau Barringer, a mining engineer, and geologist was told a story of iron chunks found around a hole in northern Arizona in 1902. In March of 1903, Barringer filed claims on the property. His four claims intersected at the center of the crater and did not extend even to the foot of the crater's slopes. He formed the Standard Iron Company in Philadelphia. But he would not see the crater for the first time until a year later. Standard Iron improved the claims before Barringer's visit. Buildings were built both in the crater and on the slopes of the rim. Shafts were dug into the crater floor as part of these early investigations by Standard Iron Company. The largest of these shafts were abandoned because of water and quicksand encountered at around 180-200 feet. Exploration continued through drilling into the crater floor. Many of the holes never reached their intended depth because of obstructions that the bit could not cut through. Tests on samples of these buried obstacles seemed to indicate that they were masses of iron meteorite metal. In some of the holes, dynamite was used to try and break up or move the obstructions. Such attempts usually failed and blocked holes would be abandoned. In total 28 holes were drilled into the floor of Meteor Crater by Standard Iron Company. Shafts and trenches were dug in many places around the exterior of the crater. Often meteorites were found in such diggings on the slopes. The floor of Meteor Crater yields no meteorites because its surface is the remains of an ancient lake. Almost everywhere else evidence was found supporting a cosmic impact origin.

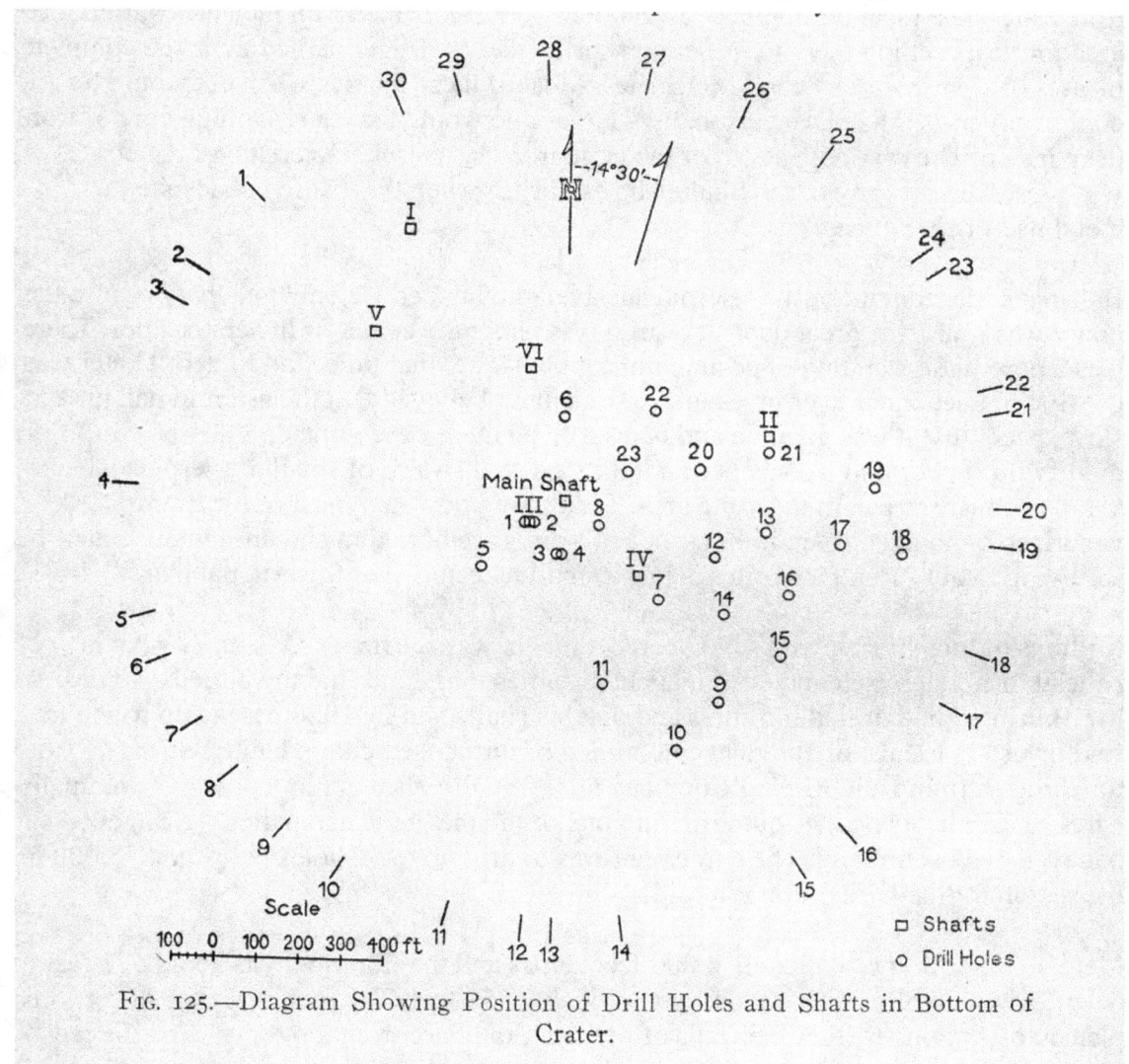

FIG. 125.—Diagram Showing Position of Drill Holes and Shafts in Bottom of Crater.

The image above is the map of the early Barringer shafts and drill holes that G. P. Merrill reproduced in his 1908 paper on Meteor Crater. The circle of numbers and lines are the stations where section information was taken. For many of the stations drawings were made to show the profile of the crater and wall at those locations. Merrill had become friends with D. M. Barringer after several years of corresponding and was an early promoter of the impact origin theory for Meteor Crater.

Most of the deeply buried meteorites recovered from 10 to 20 feet down on the slopes were in a seriously decomposed condition. The originally solid iron masses had become a hard mineralized ghost of their former state. Iron shale was the local name for the completely weathered meteorite material. This iron shale had been found in masses both large and small for decades. Thousands of fragments were lying on the surface of the ground. The large iron meteorites discovered were often covered with a layer of this iron shale. The local trading post owner, a Mr. Volz collected iron meteorites from the area for years selling them to museums worldwide. Barringer named the large complete masses of iron shale Shaleball Meteorites. Most of these masses when cut would be completely mineralized. Occasionally a larger one would have a remaining core of iron deep inside. The original pattern of the large iron crystals characteristic of meteorites was preserved in many of the Shaleball Meteorites when they were cut despite the metal itself being gone.

Barringer was convinced the asteroid survived the impact. He believed that because the crater was round the great deposit of iron was under the center of the crater floor. There was a poor understanding of cosmic impact physics at that time, and Meteor Crater was the first impact crater ever investigated. Barringer believed that the asteroid had hit at a slow speed, that it was gigantic and had survived more or less intact. Later he would modify this belief into a concept of a tightly packed swarm of smaller asteroids which simultaneously struck in the same spot. Today, we know that most of the asteroid vaporized on impact. Even if thousands of shards are buried in and around the crater the surviving total of iron meteorites is just a small percentage of the original mass.

Millions of tons of iron with seven percent nickel was Barringer's vision of what lay beneath the crater. He conducted magnetic studies to try and find the buried asteroid. Mr. Barringer had drilled his holes and dug his shafts and by 1920 was again trying to find investors to take on the financial burden of further searching. United States Smelting Refining and Mining Company after initially turning him down did eventually agree to drill holes on the south rim into one of the magnetic anomalies Barringer believed had been found. The agreement was to drill up to 10 holes or spend $75,000 in the search for the buried iron asteroid.

Meteor Crater is approximately 4,000 feet across and the floor today is about 570 feet deep. Originally the crater was deeper, probably 750 feet. The rock layers of the central plateau of Arizona seen in the walls of Meteor Crater are from top down, first the red Moenkopi Sandstone, then the cream and yellowish colored Kaibab Limestone, and finally the thick layer of grayish white Coconino Sandstone. The formations originally lay flat. Since the impact, the rocks have been thrown upward and are tilted steeply even past vertical at one area of the crater wall. There is a local faulting system that crisscrosses the region and may have caused the squarish shape of the crater, and is likely responsible for the easily recognizable fault zones in the so-called corners of the crater. The Moenkopi sandstone is visible as masses of red rock along the road leading to the crater. There is a thin layer of this distinctive red rock twenty to thirty feet thick on the top of the crater wall. The Kaibab Limestone is approximately 260 feet thick. It has the very noticeable blocky Alpha Member in the upper part of the layer. The Coconino Sandstone is the principal aquifer of the region it is 700 to 800 feet thick of which only a portion is visible in the walls of Meteor Crater. It was the much higher water table fifty millennia ago that caused the crater to flood immediately after the impact. Meteor Crater remained a lake for thousands of years slowly depositing the 90 feet of silt which forms the present surface of the crater floor.

The walls of Meteor Crater are steep. In places, it is a sheer cliff. Today there remain a handful of trails usable for researchers and staff. The public no longer has any access to the property for hiking and exploration. In 1920 numerous trails went down from the rim to the floor. Careful examination of photographs taken even today will often show the location of these fading old switchback trails. Between the flat floor of the crater and the sheer drop off of the wall is the talus zone where hiking while difficult, is a far gentler incline than the true walls. The talus is a mixture of boulders and finer sand and gravel. It is poorly consolidated rubble which has accumulated around the circumference of the pit. Gullies cut through the talus at almost regular intervals.
The slopes on the outside of the crater are mostly a gentle incline that runs down from the rim to the surrounding plain, a change of only 150 feet of elevation. The slopes are the most varied of the terrain. Some areas are nearly devoid of vegetation other parts are home to numerous scrub trees. There is a section of dunes in the southern portion of the slope. On the east and west slopes are the so-called "boulder fields" where abundant huge blocks of limestone were tossed.

This is one of the great blocks of limestone in the western boulder field. Often referred to since Barringer's time as "Whale Rock" for its resemblance to the head of those giants of the seas.

The remains of the silica mine are still very visible on the south slope. It was the most readily available and largest of the deposits of pure white silica powder found at the crater. But other deposits can be seen around the crater's perimeter. Another is on the north side just above the parking lot. Before the silica mining, which began in 1948, there had been just a few smaller extractions of silica sand from the south slope. A natural arroyo exposes the thick layer of stark white silica. The arroyo now has the remains of a road going through it. A truck loading area with a disintegrating wooden structure and the pit mine is still there as well. This silica deposit was estimated by Barringer's Company to be 2000 feet x 500 feet with a thickness of 25 feet exposed. One and one-half million tons of silica in this single location.

From the earliest days of exploration at the crater, the higher uplift of the southern rim was noted by investigators. Later referred to as the "arch" this portion of the crater was believed by Barringer to be higher because the asteroid displaced the rocks upward more above where it came to rest.

The rock thrown out of the hole by the explosion was flipped over as a flap of material. What was once above the hole now rests on the top of the rocks of the slopes, with the layers in reverse order when traced from the surface down. Some huge blocks of limestone are resting on the top of the rim while others have rolled down into the crater and can be seen on the floor today. Those on the crater floor must have fallen there recently in geological terms of time. Otherwise, they would be buried in the silty lake deposits now making up the crater floor. Some of these great blocks of limestone are 12-15 feet tall.

In the early years of the investigations at Meteor Crater researchers attempted to determine the amount of work that had been done by the impact explosion. There has never been much agreement among those making the estimates. Today the force is measured in megatons of TNT. Figures from 1.5 to 10 megatons are now the range. Though an explosive force of 15 megatons was the estimate in the 1980s. Early calculations used 300-400 million tons of rock displaced, 82 million cubic yards of rock excavated. Barringer thought that the asteroid was more than 1,000 feet across and struck at just 2 miles per second. The truth is that the asteroid was only about 150 feet across but hit at probably 8 miles per second. In 1928 Barringer believed 10 million tons of iron meteorite was buried at the crater. Modern estimates for the weight of the asteroid are from 150,000 to 300,000 tons.

It was into this confusion, lack of knowledge and heated debate that L. F. S. Holland was sent by his bosses in Boston. Like all first time viewers of Meteor Crater, it's easy to imagine him letting out a little gasp and catching his breath as he saw the chasm stretched out before him. He would climb in and out of the crater numerous times over the year of his stay. The broken, creviced, and uplifted rocks of the south rim would become a curse upon him. The high winds of the crater would plague him and his workers. On one particular day, the winds would nearly claim his life.

Meteor Crater is a majestic location. From half a mile away on the crater floor the words of the tour guide on the north rim can be easily heard. The voices of guests at the railings of the museum can be understood by someone at the center of the crater floor. The silty lake bed surface is strange to walk upon and must have also seemed that way for Mr. Holland. Today the early shafts have been filled in, fenced off or covered. In 1920 some were still open, dangerous, and Holland went down into many of them. Only Shaft No 2 is still available for researcher's use today. It has a wooden cover over the top of it. There are solar panels next to it to provide power for lights.

Several of the most visible alterations to the crater's appearance took place in the single year that Superintendent Holland was in charge. Meteor Crater was the first impact crater ever to be investigated, now well over a hundred years later it continues to yield up secrets. It has a history of not giving up those secrets easily and of frustrating investigators.

The Crater Floor

When I first arrived here, we were sleeping out in the open or in tents. I would get up early some mornings and sit on the crater edge and watch as the sun rose and the crater floor gradually emerged from the blackness. The old buildings began as just dark almost featureless spots then moments later I could see the angles and corners. Finally, I could see the various colors of their weathered wooden walls. The brown lines of rusted pipe snaked across the crater floor. Long shadows moved down the western wall and slid gradually over the floor until they were gone as the sun rose too high for any shadows to remain.

I think anyone staying here very long who believes it was made by an asteroid has wondered about what the impact day was like. The crater floor is the quietest place I have ever been. In those times when I have been alone and rested for a moment on a rock, the silence has suddenly pressed down upon me. I can close my eyes and feel the quiet surround me. Then the spell will be broken by the cry of a hawk or a sound from the distant camp. If I try hard, I can in those moments hear voices from the camp which is half a mile away. What was the sound of the impact, what does lifting millions of tons of rock and flipping it over sound like?

No one is sure when this place formed. Barringer thinks just several thousand years. Other scientists think it is much older, perhaps 20,000 years old. They cut some of the trees on the south rim and found they were 500 years old. But the crater is older than that.

I am not a superstitious man, but I have to admit that I can not be on the crater floor by myself in that silence for very long. You do not notice it when you are with someone else or are working. But if you close your eye and sit alone the silence will come and cover you.

There are no buildings on the floor of Meteor Crater today. The last part of a building to stand there intact was the annex building on the side of the main shafthouse. On April 23, 1921, a Saturday, the tremendous winds so common at the crater blew the shafthouse down. But the annex building survived. Over the previous several months the buildings on the crater floor had been stripped of their wood, and most of that lumber had been pulled up from the crater using a horse-powered winch and a wooden slide. There were several buildings in the crater. They remained from the time of D. M. Barringer's original digging and drilling work in the first years of Standard Iron Company. The crater had to be "improved" to make the claims that had been staked valid. This required digging exploration trenches and shafts all around the crater and on the inside of the pit in addition to constructing the buildings.

In late March of 1903, the first trail into the crater was cut, and development shafts began being dug. By May of 1903, the original development work was completed, and patents on the claims were applied for. The bureaucratic wheels turned as slowly in the past as they do now. It took until December for the patents on the claims to be signed. The four claims named Jupiter, Saturn, Venus, and Mars covered the floor of the crater and a portion of the slopes. The claims did not extend all the way out onto the surrounding plains. They were filed in the names of Samuel Holsinger and three of D. M. Barringer's brothers in law. Both placer mining and lode mining claims were obtained though the latter were never really considered used by the Barringers. The patents on the claims were signed on December 24, 1903, by President Theodore Roosevelt, Barringer's friend. In March of 1904, D. M. Barringer made his first visit to the crater. Accompanied by Benjamin Chew Tilghman, his partner in Standard Iron Company until 1910 when he withdrew from the company. Tilghman over the years became increasingly discouraged about the possibility of finding the asteroid by drilling into the crater floor.

They arrived in March, and by April of 1904, the first shaft was being dug at the center of the crater. The lake deposits covering the crater bottom were discovered during the digging of this first shaft. It appears that immediately after the rocks were penetrated by the asteroid the thick layer of Coconino Sandstone, the aquifer of the region partially filled the crater with water. It remained a lake for thousands of years.

Shells like the one above of aquatic animals can be found on the crater floor still today testimony to the thousands of years the basin was a lake.
One of the layers discovered in the digging of the first shaft was a zone of pumiceous

rocks above the regular mixed fallback rubble from the impact. These light, fluffy rocks that were created by the impact floated for a time on the surface of the water pouring in from the Coconino Sandstone. After some time the pumice-like rocks did become waterlogged and sank. They formed a layer devoid of other rocks just below the silty lakebed deposits that continued to accumulate for thousands of years. Approximately ninety feet of lake sediments were found by the miners digging the first shaft. The digging was easy, and in a very short time, they were at 181 feet of depth. It was here the water and quicksand were encountered. The shaft was abandoned.

Work outside the crater on the slopes continued, and many small iron meteorites were gathered from the surface. The 225-pound mass was also found during this early phase of exploration. Shaleball masses of the weathered iron meteorite were found in the trenches dug at various locations. But in the interior of the crater, the focus had turned to drilling exploratory holes. Two were drilled in 1904, and three more in the spring of 1905. These five constitute a group unto themselves and were separate from the later holes drilled. Only hole number five reached below the crater to undamaged rock. Drilling on it was stopped at 1003 feet. The other four holes all hit obstructions or had other difficulties that forced them to be abandoned. Metallic material was brought up from some of the holes. It was clear that the obstructions were often masses of iron meteorite. They could not be drilled through or bypassed. The water in the drill holes would quickly turn green and become foul. The material brought up would respond to a magnet and test positive for nickel. Sometimes dynamite was placed in the holes in an attempt to break up or move obstructions. This effort was fruitless. The obstruction in one case only moved a few inches. In another of the holes, the drill string weighing 1800 pounds was dropped onto the obstruction, and a loud metallic ring was heard each time it struck whatever was at the bottom.

The casing was abandoned in not just these first five holes but in others to follow. This iron will become an issue for Mr. Holland and Mr. Fay during the magnetic surveys. Water tanks were built in the crater and on the rim. Pipelines were built into the crater and ran along the walls and across the floor. All this iron confused the compass during the magnetic survey work inside the crater in 1920.

In July of 1905 work began on another shaft at the crater center. This was a much more elaborate operation. A steam hoist and of course the boiler to make the steam were lowered into the crater in pieces and assembled on the floor. The machinery rests on large foundations; they are obvious remnants today. More large iron objects to make magnetic readings erratic in 1920. The second shaft was a larger two compartment shaft. By August after just a month of digging it is at 100 feet in depth. Barringer and Tilghman were making plans for the attack on the quicksand layer. The scheme they had in mind was to work in continuous shifts and do constant hoisting while steam-powered pumps removed the water. By December this plan was in operation, and they were struggling to push through the water and quicksand. January 1906 found the miners still fighting to reach 200 feet in depth. They brought in more equipment, and the shafthouse was build in the center of the crater. On January 19 mining engineer Neville telegraphed Barringer informing him there was no possible way to continue digging deeper. The timbers had nothing to hold onto, and they were digging in mud.

Barringer will not order this shaft abandoned until March 26, 1906. At which time the miners were dismissed and work stopped after spending $55,000. The shafthouse will stand until that fateful Saturday of April 23, 1921, when most of the building collapses. The shaft was used periodically to check the level of the water in the area. By 1921 the shaft was nearly dry with just two inches of very foul stinking water measured at the bottom by Mr. Holland. The pollution of the water was likely because of animals falling into the shaft and dying. There was scrap iron in this shaft as well when the magnetic survey was done. Barringer was not aware metal had been dumped down this shaft. Holland was instructed to have Mr. Fay go down into the shaft and take a magnetic declination reading. Barringer took a reading about 15 years before with a crude instrument and read a slight indication of a different magnetic declination. He had long desired that another reading be taken with better equipment. In the end, the instrument used was just a compass and not a dip needle. All the metal dumped down the shaft since Barringer's early reading made getting an accurate reading impossible for Mr. Fay in 1920.

Today the old shafts in the center of the crater are surrounded by tall chain link fences with three lines of barbed wire on top. They are open holes left for future research perhaps or because it was too expensive and too much trouble to fill these larger ones. Holland and a miner were in the center shaft only hours before the shafthouse blew down. It is rumored that some of the metal parts from the Cessna airplane which crashed in the crater August 8, 1964, were put into one of the original shafts. Pieces of the plane's fuselage remained on the edge of the talus on the western side of the crater floor.

By the end of the early years of Barringer's work, additional shafts will be dug into the crater floor. Of these only one is preserved for use. It is off to the east slightly from the center of the crater. It is covered with a wooden lid, and solar panels are set up next to it to power the lighting in the shaft or more likely to charge batteries that supply power. The only evidence left of the other shafts are shallow depressions in the surface where the backfilling of the holes has subsided, or low mounds of bare ground where the dug up material was spread.

While these additional shafts were being dug more holes were also being drilled. Three holes were drilled in 1906. After the failure of the larger shaft at the crater center, drilling escalated. In 1907 sixteen holes were drilled. In 1908 four more were drilled bringing the total to 28 holes in the crater. Number 28 was abandoned in July of 1908. The last 23 of these holes were drilled in a small area of the crater center. In actuality, only one-twentieth of the floor was surveyed by drilling. In August 1908 Barringer

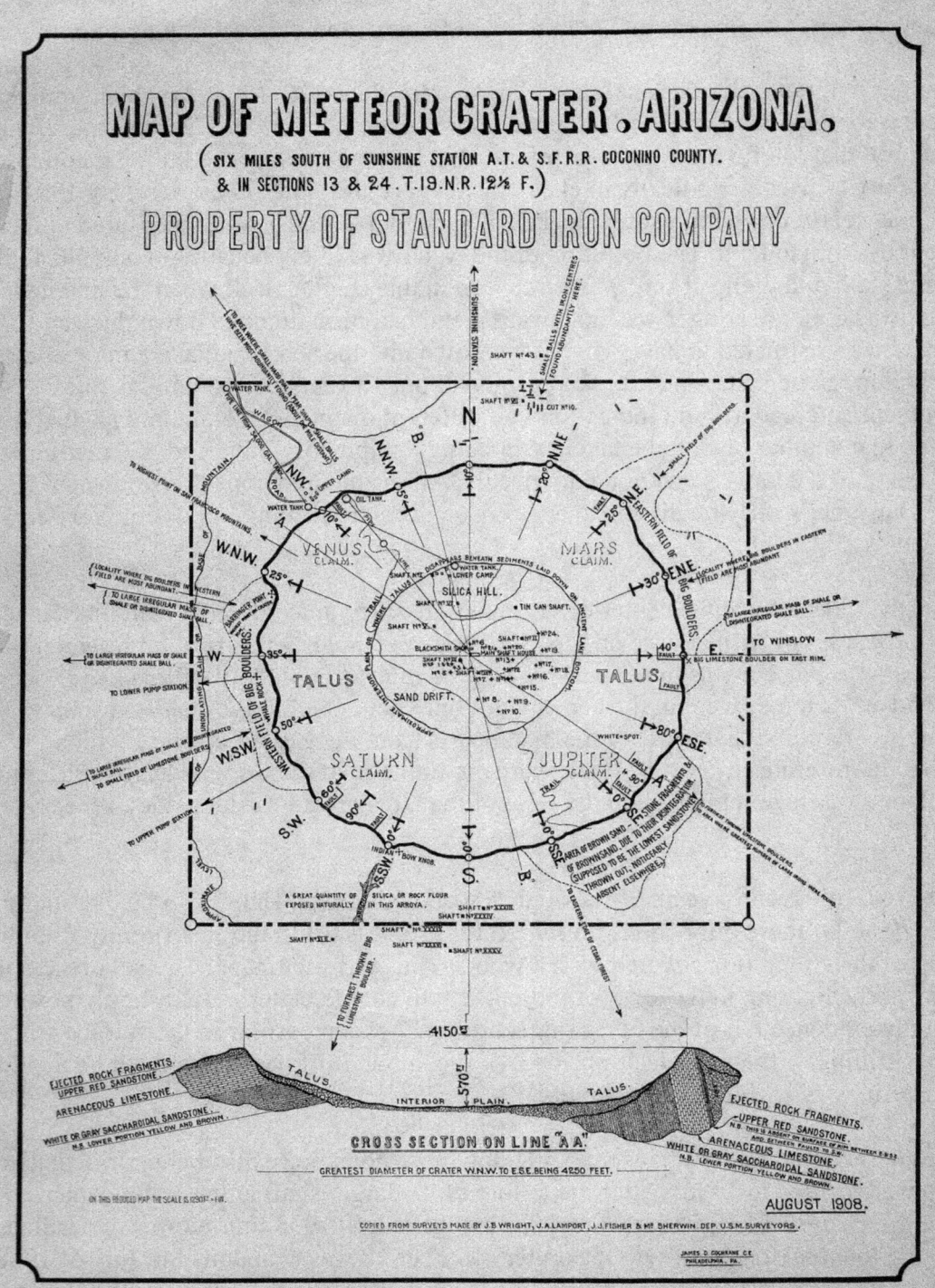

orders that all drilling be stopped. There is by this time much iron left in holes or scattered around the floor. Old machinery such as the remains of drill rigs and pumps can be found on the floor still today resting where they were abandoned or dumped over a hundred years ago. Mr. Holland will have to deal with the materials on the crater floor. He will salvage whatever useable pipe he can, and he will have men strip the old buildings of lumber. He will haul these items out of the crater. A 1200 foot long slide was built from this wood. Later the slide will itself be cannibalized. Its wood will be used in the tunnel driven into the south wall in the spring and summer of 1921.

On the eastern side of the crater floor, there are many locations where machine parts from drilling equipment such as pictured here can be seen lying where they were discarded over a hundred years ago.

All the abandoned metal on the floor of Meteor Crater will be dismissed along with the erratic readings it causes in the magnetic survey. The survey is to locate the variations in the local magnetic field and determine sites for larger churn drill holes on the crater rim. Since all the metal on the crater floor is more than a thousand feet away and scattered about, it is decided that it does not affect the magnetic results gotten on the crater rim and slope.

None of the buildings remain today on the floor. Except for the bits of trash and waste in the form of rusty cans, broken glass and crockery visible where the buildings once stood, there is little evidence buildings were ever there. Besides three of the shafts, the boiler and winch nothing is visible from the top of the rim today. Hiking on the rim and into the crater stopped decades ago. Access to the interior of the crater is strictly controlled. Though occasionally someone will find their way into the crater and have to be removed. Water bottles are left on the foundation of the boiler for the safety of such intruders. It can be dangerously hot on the crater floor, or bitterly cold. The harsh cold is what a man in 2013 experienced. He climbed over the fence and barbed wire surrounding an old Barringer shaft in early January. In an effort to "appease the gods" he jumped feet first down the shaft. He was rescued in a herculean effort by local agencies. He suffered broken bones and other injuries in the fall. He was very fortunate that someone saw him jump.

As a laboratory for research Meteor Crater remains the best-preserved impact crater on Earth. A hundred years ago the crater was a beehive of activity as Standard Iron Company attempted to extract valuable resources from the site. There were some years after this initial burst of effort when very little work was done. Then in the spring of 1920, Mr. Holland arrives and begins the work of drilling to the buried asteroid from the south rim. Once again Meteor Crater is buzzing with activity.

The Meteorites

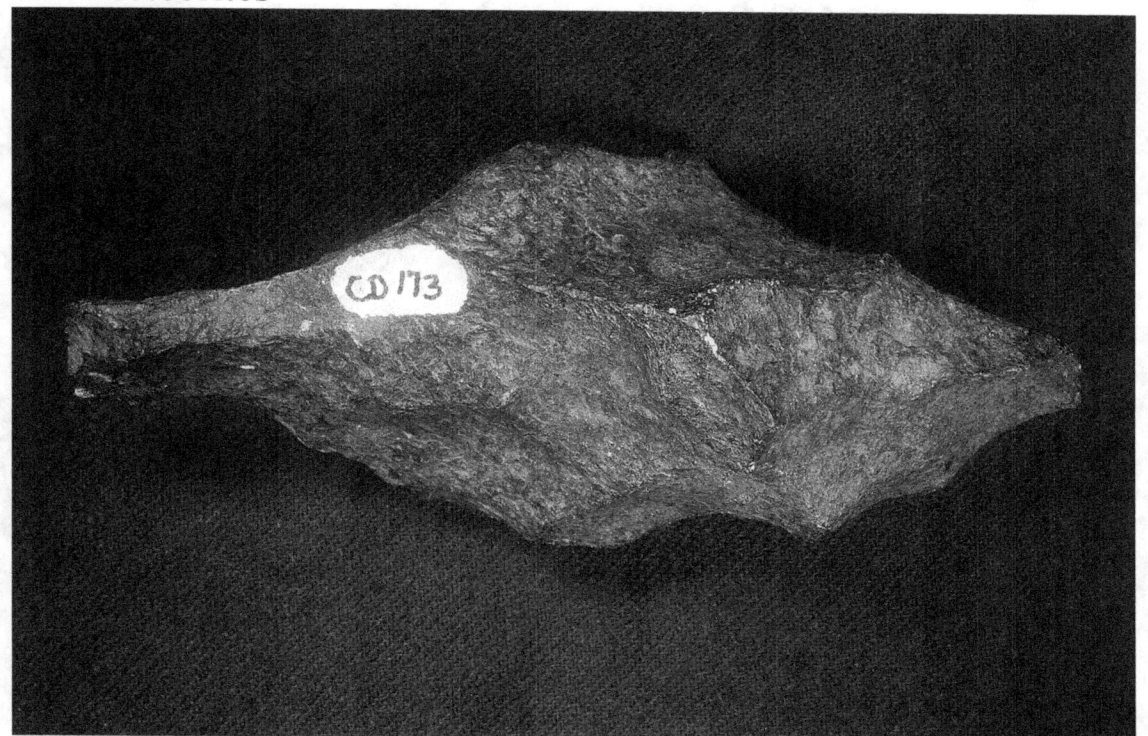

Everyday bits of material from space strike the Earth's atmosphere and are seen as streaks of light which quickly extinguish. These are normal shooting stars. They are tiny bits of comet dust and rocks as small as sand and only as large as a grain of rice. On much rarer occasions a larger piece of material from space will enter the atmosphere and survive to reach the ground. These pieces of space rock when recovered are meteorites. They are pieces broken off larger asteroids far out in the solar system between Mars and Jupiter. They have been flung into orbits which have sent them into the inner solar system to wander until they are on a collision course with Earth. Meteorites tell scientists the story of the solar system's history. Meteorites have been grouped into three main families, the stony meteorites, the iron meteorites and the stony-iron meteorites. The asteroid which formed Meteor Crater was composed of nearly solid nickel-iron metal. It shattered on impact creating thousands of fragments that landed in the area of a few miles around the crater. The meteorite shown above is one such piece. The meteorites found near Meteor Crater are called Canyon Diablo meteorites. Canyon Diablo is the winding canyon and seasonal stream to the west of the crater. It was not until years after the crater's discovery that it received an official name. The long debate about the crater's origin and whether or not the meteorites were even related to the crater led to the meteorites being named after the canyon the nearest geographical feature with an established name. That continues to be the naming method in the United States for most meteorites. They are named for the nearest city, mountain, dry lake or similar feature.

Most of the asteroid's 150,000 tons or more were vaporized and became tiny iron droplets a few trillion of which are still present in the soil. A sample containing approximately a thousand of these tiny nickel-iron droplets is shown in the next photograph along with a sewing needle for size comparison.

Canyon Diablo iron meteorites are members of a smaller family within the irons called the coarse octahedrite group. As the metal alloy of the asteroid's parent body cooled over millions of years in space, the nickel-iron crystallized to form a structure that is octahedral in shape. The amount of nickel in the bulk alloy of the asteroid determined the size and number of crystals that could form of two nickel-iron minerals. When the bulk alloy contains a generous but not excessive amount of nickel, numerous small crystals form. If there was less nickel in the original molten metal then fewer but larger crystals formed. The nickel-rich mineral Taenite forms as long as there is sufficient nickel. But when the amount decreases because of crystal growth it becomes too low for taenite to form any longer. Then the low nickel mineral Kamacite only crystallizes. If too much nickel is in the bulk alloy only taenite crystals form. These nickel-rich meteorites are of another small group called the Ataxites. If there is very low nickel content in the original molten alloy, then Taenite never forms, and only Kamacite crystals grow. These nickel-poor meteorites are called Hexahedrites. The important thing is that all iron meteorites contain nickel. Many times a rusted cannonball or mill ball from a rock crusher will be thought to be a meteorite by the finder. To their dismay, they find out later when the metal is tested that there is no nickel present.

Octahedrite meteorites like Canyon Diablo are a mixture of both nickel-iron minerals and have a pattern of crystal growth that can be seen when the meteorite is cut, polished and etched with acid. The observed structure of crystals is called the Widmanstätten pattern.

Canyon Diablo meteorites with 7% nickel have quite large crystals of the two nickel-iron minerals thus the name coarse octahedrite. Mixed into the Meteor Crater meteorites are trace amounts of many other metals and elements. Iron meteorites often have masses of graphite and other minerals within them. These associate minerals constitute a small portion of the Canyon Diablo meteorite composition. In some iron meteorites, these other minerals are a much larger component of the mass. Graphite is a soft mineral of carbon and is found mainly as roundish nodules in Canyon Diablo meteorites. Graphite is a mineral commonly mined on Earth. It consists of crystal layers that lay flat on top of each other like sheets of paper in a ream package. This gives graphite useful properties for many products including writing pencils and lubricants. Some Canyon Diablo meteorites contain another form of carbon. Diamonds were discovered almost as soon as the Canyon Diablo meteorites were first examined. The grinding wheels used to smooth cut surfaces of meteorite would be deeply grooved when grinding them. Upon inspection, small diamond crystals were seen in the meteorites. It was later recognized that not all Canyon Diablo meteorites showed the presence of diamonds. Diamonds are only found in the fragments from the slopes of the crater. These diamond bearing meteorites also had other evidence that indicated they were exposed to great pressure and temperature. The heat of the impact damaged the Widmanstätten pattern of these meteorites. The presence of the diamonds indicated that enormous shock and pressure had gone through the metal converting in an instant some of the graphite into diamond. Today most scientists accept that the diamonds found in Canyon Diablo meteorites formed in the impact. The meteorite fragments found on the plains surrounding the crater do not have diamonds or damaged Widmanstätten patterns. A small number of meteorite types do contain tiny diamonds that formed while they were out in space. But none of these types are iron meteorites.

Barringer sent meteorite samples repeatedly to laboratories and smelters trying to learn what they were made of. Analysis of metals 100 years ago was far different from the processes done today. Today, we could place a sample in front of a handheld X-ray fluorescence (XRF) analyzer and get immediate results for the elements contained in the metal. In 1905 Henri Moissan dissolved 116 pounds of Canyon Diablo meteorite in hydrochloric acid and then tested the resulting solution with reagents. To just make a cut through a Canyon Diablo meteorite 10 inches by 10 inches took 20 days for him to accomplish. Moissan is credited with the first discovery of carbon silicide in nature. He found it in a meteorite. He would also discover a crystalline mineral in meteorites later named for him; the gemstone Moissanite is today lab created for jewelry.

Barringer had provided United States Smelting Refining and Mining Co. with a supply of both meteorite and iron shale for analysis. They received this material when negotiations had first begun in 1918. Though these discussions did not lead to an agreement at that time the analysis of the meteoric material was conducted. Later Mr. Roy B. Earling of USSR&M Co sent a copy of this analysis for "platinoid metals" to Mr. Holland. Barringer had said these valuable metals were mixed in with the nickel and iron. The analysis showed .285 ounces of Platinum per ton of meteorite iron. This was just 8.079 grams per ton. For comparison, the US five-cent coin, a nickel, weighs 5 grams. The largest iron meteorite ever found near the crater weighed far less than a ton. It would have yielded something around five grams of Platinum when smelted. The amount of Platinum found in the analysis of the iron shale was just 3.74 grams per ton. The amount of Palladium 1.8 grams per ton in the iron meteorite material was just about half the amount of Platinum. But the Palladium in the shaleball material was double that found in the iron meteorites at 2.863 grams per ton. It is reasonable to believe that the iron shale being far lighter in weight than the iron meteorites took much more volume to make a ton. While the iron and nickel were found to have leached out of the iron shale, the Palladium appears to have stayed in the weathered byproduct.

While all this was interesting a hundred years ago for men in the metals business, it is the statement by Mr. Schleicher, the chemist about problems he had shaving the metal in preparation for testing which is somewhat more interesting today. He had some difficulties with diamonds in the remaining iron center of a shaleball specimen. He shaved the iron meteorites and the metallic cores of shaleballs up on a shaving machine into the small bits needed for testing. Presumably, he dissolved these shavings in acid for the chemical analysis. Here is what he said about the one shaleball that he could not shave more than 2/3. "Due not only to the hard nature of the iron but more to the loose gritty constituent the tools would not last a single stroke across the piece. Several tools were tried, even some from the machinists who had tools of their own which would cut anything! (They didn't cut the shale iron). Stellite was tried with disastrous results. The remaining third of the piece was broken into several smaller pieces with a hammer."

The large piece of iron shale shown here demonstrates what was first noticed by the early investigators. It forms as thick curved layers on the outside of large iron meteorites during weathering. Gradually the chunks exfoliate off the meteorites and are scattered. Small pieces will have a thick chunky appearance larger pieces will have curved shapes. Over the thousands of years since the impact, the meteorites exposed to water have been consumed by the process of creating the iron shale. Nothing is left now of these meteorites, but hundreds or thousands of small bits of the hard iron oxide rock.

Diamonds were being used in industrial applications already in the 1920s. United States Stores, another of the subsidiaries of USSR&M Co. sent a letter to Mr. Holland of Crater Mining Co. and three other companies inquiring if they wished to purchase black diamonds. They could be mounted in bar stock to be used as grinding wheel dressers or mounted in bits for cutters. Mr. Holland had no use for them with a churn drill. But the diamonds in tons of Canyon Diablo meteorite might have added to the value of the metal. The price offered to Holland was $135 per carat for 50 carats or less. $120 per carat for 100 carats, $112 per carat for 200 carats and $110 per carat if 300 carats were purchased. Diamonds might have been the easiest of the rare matter to extract from the metal. The platinoid metals as Barringer called them would not be easy to recover. Barringer had approached Johnson Matthey & Company who specialized in precious metals, about investing in exploration at the crater. They turned him down being unable to figure out how they would get the tiny amounts of the precious metals out of the nickel-iron profitable. Johnson Matthey Co. was founded in 1817 and is still in business, still a world leader in precious metals.

Barringer was excited when told that the meteorites contained 14.96 grams of iridium and over 3 grams of platinum per ton. Even such small amounts of these rare metals would double the value of the iron he hoped to find. He made the following statement. "If that stuff is worth $200 per ton I will make the interior of the crater look like a rabbit warren before I give up my search for the projectile that certainly caused the hole." He did drill 28 holes into the crater floor. He also dug many shafts in and around the crater searching for his projectile. After little success searching on the floor of the crater, he moved his attention to finding the asteroid under the south rim. Iron meteorites are strongly affected by magnets. It seemed reasonable at that time that magnets would be likewise affected by a huge buried iron mass. He had magnetic surveys of the crater and the surrounding area made to determine if there were any local deviations from the Earth's natural magnet field.

One gram of iridium is seen in the image above. The cube of iron meteorite shown is one centimeter on a side. Iridium has become an indicator of asteroid impacts since Barringer's time. It is so rare on Earth that where it is found in any appreciable amount, it is because it was deposited there by the impact of an asteroid. There were only a few uses for iridium in 1920, but there are many more today, it has become an important rare metal.

The largest meteorite fragment was found in 1911 one-and-a-half miles northeast of the crater. It was discovered by Samuel Holsinger. He was Barringer's foreman of sorts at the crater in the early years. It had been Holsinger who first told Barringer of the iron deposit in 1902. Barringer was in England at the time the 1400 pound fragment was discovered. They were building the first museum on the north rim and Barringer said to put the specimen there. Holsinger would die in August of 1911 before the building was completed. Barringer expressed a wish that the large piece of the asteroid always be called the Holsinger Fragment. The meteorite remains on display in the modern museum on the crater rim. It was housed in the original stone masonry museum for several decades. The story is told that a small hammer was placed next to it so that visitors could strike the meteorite to hear the metal ring. A fire consumed the old museum, but its ruins remain on the northwest rim.

The Holsinger Meteorite is the largest discovered fragment of the 150-foot (45-meter) meteor that created Meteor Crater.

By the time that Barringer filed his claims on just the crater itself, the trading post owner at Canyon Diablo Station had long been collecting and selling the iron masses to museums all over the world. His name was Fred W. Volz, and he estimated 10,000 pounds of iron meteorites were found and sold. Thousands of Canyon Diablo meteorites have been recovered since the first was found prior to 1891. From a fraction of a gram to hundreds of pounds Canyon Diablo meteorites come in all sizes and shapes. So effective were Volz's efforts that he created a glut of Canyon Diablos on the world market. No large museum was interested in obtaining them any longer.

CABLE ADDRESS "MUSEOLOGY NEW YORK"

AMERICAN MUSEUM OF NATURAL HISTORY
77TH STREET AND CENTRAL PARK WEST
NEW YORK

DEPARTMENT OF GEOLOGY AND
INVERTEBRATE PALÆONTOLOGY
EDMUND OTIS HOVEY, PH.D., CURATOR

29 November, 1920.

Dear Sir:

Replying to your esteemed favor of the 21st instant I would say that inasmuch as we have many of the so-called Canyon Diablo meteorites the acquisition of additional specimens would not be of interest to us, unless they showed some new peculiarity. If, however, a mass weighing four to five thousand pounds should come to light we should be glad to enter into negotiations for its purchase. We should certainly be willing to pay much more for it than its value as ore could amount to.

Yours very truly,

E. O. Hovey
Curator.

Mr. L. F. Holland,
 The Crater Mining Co.,
  WINSLOW, Arizona.

EOH/EGM

When Mr. Holland wrote to the American Museum of Natural History on November 21, 1920, offering meteorites from the crater he quickly receives back an answer of no. The curator of the AMNH in New York was Edmund Otis Hovey, a well-respected geologist. He mentions in his reply letter to Holland on November 29 that they have many Canyon Diablo meteorites and would not be interested in more unless they were to show some newly found peculiarity. He adds however that they would enter into negotiations for any meteorite mass of four to five thousand pounds that might be found. He says they would "be willing to pay much more for it than its value as ore could amount to." Hovey was not unfamiliar with meteorites. He had traveled to Portland, Oregon in 1906 to inspect the 15.5 ton Willamette meteorite which would be purchased the same year becoming part of the American Museum of Natural History collection.

Barringer's mining claims were just for the area of the crater's interior and a portion of the slopes. He had no claims in the region around the crater. The meteorites collected miles from the crater including the Holsinger Fragment were taken from a property where Barringer had no rights. In the 1920s the friendly relationship and helpfulness of the local rancher, a Mr. Hart may indicate that there was some form of an agreement for land use and the meteorites found. Perhaps Mr. Hart was to participate financially in some way after the buried asteroid was located? This was never discussed in the Holland papers. No meteorites are found near the surface inside the crater. For the most part, the meteorites from Barringer's portion of the slopes were small when recovered on the surface and weathered badly when dug up. Barringer mapped the location of hundreds of meteorites and iron shale bits that were found out on the surrounding plain. He printed the maps in the scientific papers he wrote on the crater. He was doing nothing other than Volz had been doing. He may have believed he owned the crater and the crater was the source of the meteorites. Therefore he might have believed he had a right to collect them. Sort of like the owner of the highest point on a buried vein of ore can follow all the gold below as long as the vein continues regardless of who owns the property at the surface. Barringer took meteorites where ever they were found. Maybe in total disregard for who owned the property. The owners of the meteorites in many cases were likely the people of the Arizona Territory which in 1912 became the people of the State of Arizona. The property around the crater is presently a checkerboard of square parcels belonging alternately to the state and the local ranch. The fact that much of the property is still state-owned would suggest that all or most of the land might have been so a hundred years ago as well.

It is interesting to note on the Barringer map shown above that seven meteorites of between 10 pounds and 1000 pounds were found in the cedar forest three and one-half miles to the south-east of the crater by Volz's hired men before "Acquisition of Property by S.I. Co." meaning Standard Iron Company or specifically Barringer. The map is signed by S. J. Holsinger and it is likely he designed the map, but the idea that they had acquired searching rights to a larger surrounding area is one which will be stated by Barringer on many occasions using the same "acquired the property" statement for land he never officially acquired at all.

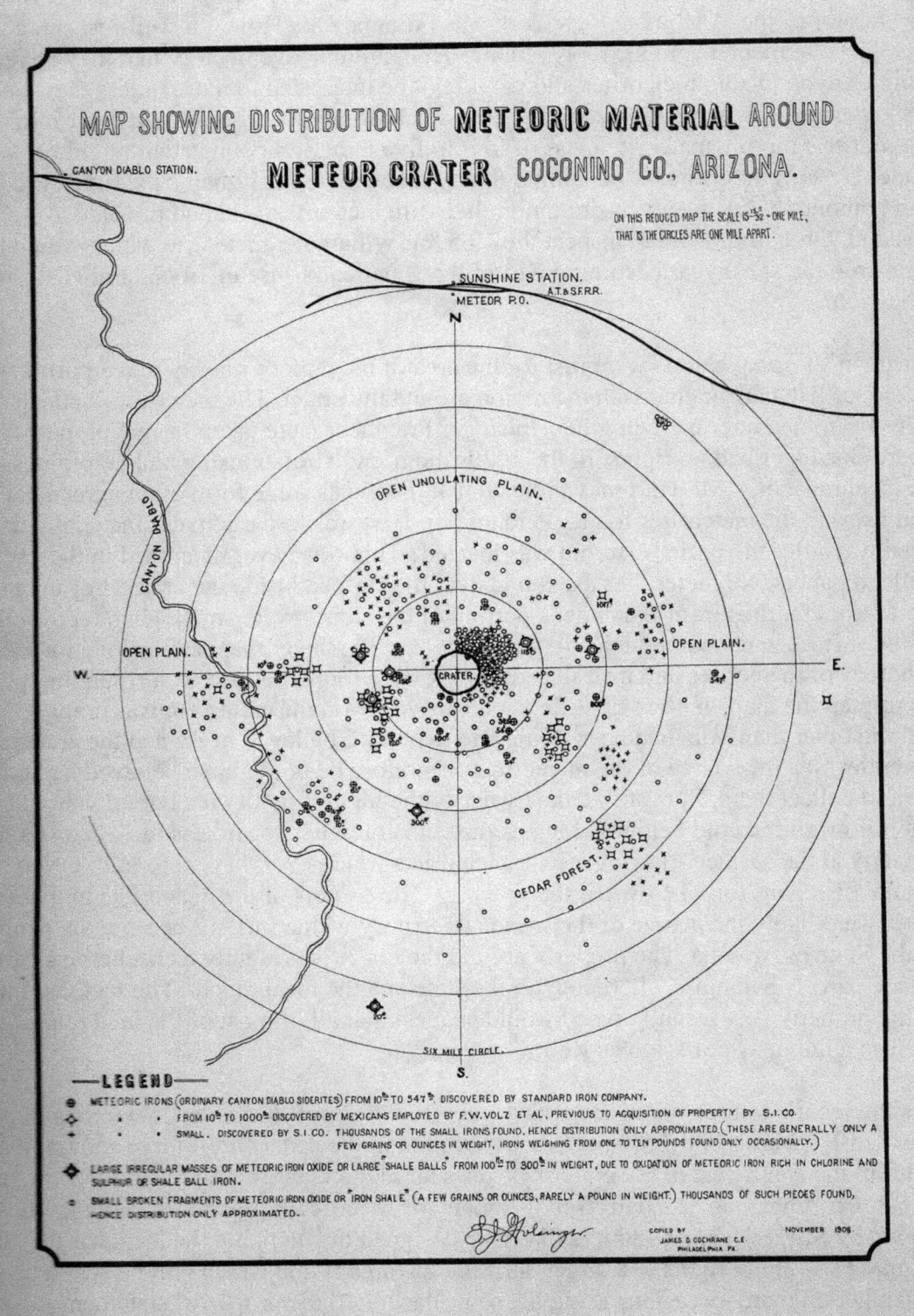

As the decades pass more and more official permission will be required to hunt the area around the crater for meteorite material. Today no hunting of meteorites is allowed. Signs warn of prosecution for trespassing and the taking of meteoritic material. There is something to be said for leaving the meteorites buried. After 50,000 years they have weathered about as much as they will. Their further deterioration will be very slow indeed. They have developed layers of magnetite and other minerals on their surface which helps to protect them from further rusting. With all the collecting for over a century, any patterns of distribution from the impact have probably been erased. With any historical site, it is good if possible to leave a portion undug for the future when better archeological tools and practices have developed.

The evidence of the impact that does remain intact is found in the tiny droplets of nickel-iron present in the soil around the crater. These "spheroids" as Dr. Nininger named them amount to thousands of tons of remaining asteroid material. They represent mixtures of metals from the original meteorite that were sufficiently resistant to oxidation to survive until the present. Nininger spent years living near the crater doing research. The analysis of the spheroids he collected from the soil showed they averaged twice or more the amount of nickel and cobalt found in the meteorite fragments. Cooling as they did from the vapor cloud of the impact they have a wide range of composition. Some large portion of the spheroids undoubtedly had less nickel and cobalt. Lacking the corrosion protection the nickel supplied these deteriorated to nothing long ago. Only around 30 tons of iron meteorite fragments have been found at Meteor Crater. As much as a thousand times that amount is represented by the quantity and composition of the spheroids surviving in the soil.

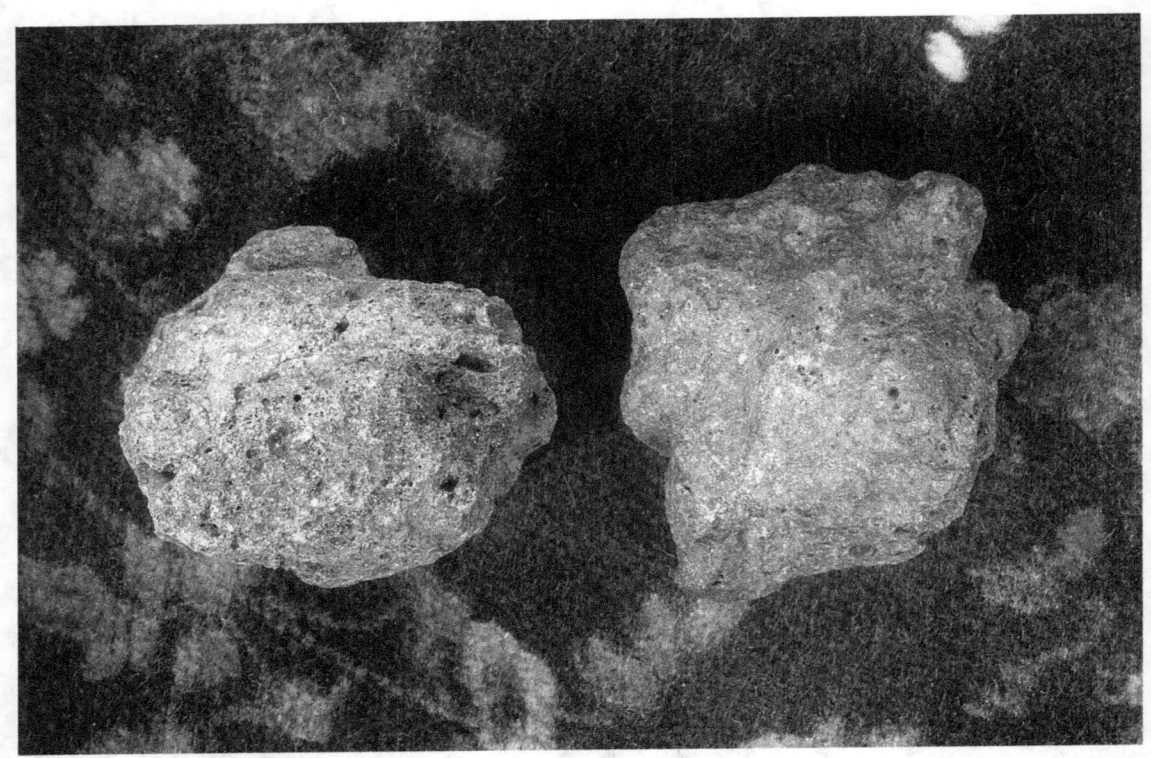

At many impact craters, melted rock products often referred to as impactites are recovered. It was many decades after the first investigations at Meteor Crater before any impactites were found, But in the 1940s Dr. Nininger finally discovered them at several locations around the crater. They are a porous poorly melted clinker type rock rarely the size of a walnut and usually much smaller. Many of these impactites will hold a magnet since they contain melted blebs of nickel-iron. Even the small percentage that will not hold a magnet will show tiny shiny specks of metal when cut. Since the metal is completely melted it is theorized that the impactites formed during the same phase of the crater's formation as the tiny melted iron spheroids. That some of the spheroids were mixed into the crushed rock which was melted and blown up into the air becoming porous glassy blobs. The image above is of two Meteor Crater impactites. Impactites from Meteor Crater have several different types of rock fragments incorporated into the porous melt. Bits of Kaibab Limestone, Coconino Sandstone, Moencopi Sandstone, are mixed with the glassy material.

It was not long after the discovery of the impactites that a much tinnier glassy product was recognized at Meteor Crater. Dr. Nininger named these grain sized objects "bomblets" and he found thousands of them. They occurred in the soil of the crater slopes and more thinly in samplings made around the crater. Made of melted rock that formed droplets some of these will respond to a magnet and occasionally a tiny iron spheroid will be found adhering to the surface of a bomblet. It was proposed that these formed moments after the tremendous heat vaporized the crushed native rocks and iron meteorite. The rock cooled and became droplets just as the iron had which formed the spheroids. Bomblets occur in many shapes some are just broken glassy bits, while others are smooth round or ovoid shapes. They range in size from .6 millimeter to 2 millimeters. The average weight of bomblets is just 1.7 milligrams. A group of these objects is shown below.

Other fragments of microscopic size were found early in the investigations and add further to the surviving amount of material. These bits have the same composition as the iron meteorites and represent small pieces of the asteroid that were shredded and torn off during the impact but not vaporized. Along with these are the pieces of iron shale which were so common everywhere a hundred years ago. Great quantities of iron shale have been collected. It is still present in the soil as tiny bits and larger chunks for miles around, but it is not seen everywhere any longer as it was when the crater was first discovered. There never was 10 million tons of nickel-iron, but scientists can account for much of the asteroid's actual 150,000 tons.

The People

L.F.S. Holland

There was no weather or traffic report, certainly no good morning program on television. Only an occasional newspaper might float around the job site. No telephone was there; even a telegraph was many miles away. The following might have been Superintendent Holland's thoughts in 1920 as the eastern horizon slowly brightened and another hard day was beginning.

I'm a mining superintendent in the middle of Arizona. In fact, as the sun is rising I am waking for the day in a wooden building with a black iron stove for heat. The walls are lined with building paper to keep out the cold. But it is still cold. I slept alright on the cot and rather thin mattress. But anyone would sleep after the days we've been working for three months. I'm used to living in mining camps. The rugged surroundings of Meteor Crater with its lack of conveniences are neither better nor worse than at other places I've worked; places as far away as Nova Scotia. I hired a cook who is reported to be the best in this part of Arizona and conditions are better now that the camp is complete. At least we are not sleeping in the open anymore, and the meals are not prepared in a tent. But on mornings as cold as this one, I long for home in Hollywood.

All the workers who have shivered through the cold night are pulling on their work clothes and fastening up their boots. The drill crew from California is made of good men who work hard. They get paid well too. But the casual laborers who finished the last of their jobs in November were another story. It seemed impossible to get men to work for very long, especially if they had to hike in and out of the crater. I only found one man yesterday in Flagstaff to hire for odd jobs, and he looks to be of little account.

In the winter the reservoir and pipes can freeze, and work must stop. Snow often covers the crater the locals say. The men during those times face the prospect of being laid off until it thaws. But nothing is frozen today even though it is early February and cold. It has been too dry as well. Not nearly enough rain. Storms have been all around the area, but none have benefited the crater's water supply. The reservoirs out at the canyon do not have enough water to supply a long drilling program. We had just the one little rain back in October. Is that right? Yes, October, it was when the lightning struck the museum and put a hole in the roof. Since then no rainwater for the reservoirs. August was the last good rain when the dams leaked so badly.

The broken drill bit is still lying flat in two pieces at the bottom of the hole. I am walking the short distance to the drill site high on the south rim of Meteor Crater. It is nearly 8 AM, and the daylight tower will start. That is the one I have been supervising, but I'm often up there during the other two shifts. I persuaded the drilling foreman to work the evening tower and have had to leave the morning shift without a supervisor. There are usually just two or three men on each shift anyway.

We will try again with all our strength and experience to recover the underreamer and get back to drilling down. I could see in the men's faces as they ate their breakfast and drank their coffee that everyone knew the reasons for the current struggle that is now nearly two weeks on. I share the despair they feel about achieving success. But it's not the first serious problem. There has been so much breaking of tools. In one week four stems broke, we never recovered some of the parts. The day they brought the war-years drill stems I thought there might be problems. They are too light, and the steel is not strong, the casing provided is little better. The gasoline engine has never been quite right though it is better now than at first. The rock is so fractured that we can not get the drill to run straight. It follows every crevice in the rock. The first of November we spudded in and drilled down 97 feet in less than three days. Drilling on three towers with high hopes, but no longer, only problems now. There were plenty of crevices even at the beginning. But we are down a long ways more, and the fissures continue to be everywhere. They are often filled with gravel and silica powder. The drill bit wants to follow that soft stuff. We put water down the hole, and it runs away out the bottom leaving us no way to use the bailer. I sent a lantern down the hole, and its light disappeared completely into one of the cavernous fissures.

The contract is to drill up to 10 holes or spend $75,000 to find the iron meteorite. We had spent a little over $60,000 by the time we started this first hole. This work at Meteor Crater has more challenges and problems than any job I can remember. So often now I wish I was back in Cherry Creek, Nevada or Telluride, Colorado, anywhere else from my past.

I pull my coat around me; the morning chill is still in the air, but thankfully the wind is down today. It roars so hard sometimes here on the rim I am convinced it will blow me and the rig away. The old shafthouse on the crater floor sings like a concertina as its walls sway back and forth. The sky is free of clouds this morning. The color is the same as the turquoise jewelry I saw at Volz's Trading Post. But that was when I arrived, how long ago that seems now. They are serious about digging a shaft into the side of the crater to get the drill tools and clear the hole. That is if we can't bring them up soon. They want me to have the surveyor locate the position for the tunnel. I think it will cost me my job if they decide to dig. They have to be concerned about tunneling. They know how broken up the rock is. Barringer told them it is dangerous. Three months of work and we are at just 312 feet. They planned for a 1,000-foot hole or more. If they dig that tunnel to remove the tools, it will be at least a couple months more before we get going again. Will they keep me on after that? Do I even want to stay here anymore? Times are tough in the mining industry and a job, any job is a good thing.

The sun is up, and I'm standing again on the edge of the rim. The great crater still in long shadows is laid out in front of me. These racing thoughts have to step back so my mind can focus on today. The morning tower is leaving for breakfast and sleep. The men with me are ready for more fishing to clear the hole.

Laurier Fox Strangways Holland was born in London, England in 1871. He uses his initials exclusively in all correspondence. His permanent residence in 1920 is 1718 La Brea Avenue, Hollywood, California. The Los Angeles area near him is in an oil boom; derricks are a veritable forest across the land. He is familiar with drilling though much of his work has been in mine exploration and mine engineering. He was the manager of the Evangeline Gold Mine in Nova Scotia and responsible for reopening that mine in 1907. He and his wife were attendees at the ninety-third meeting of the American Institute of Mining Engineers held in Toronto, Canada the same year. They enjoyed the train trip to Cobalt, Canada as part of the activities of the meeting. They were still using Telluride, Colorado as their address while in Nova Scotia. He will return to work in the Telluride area after leaving Canada. Holland managed the Smuggler-Union Mine in the San Juan Mountains near Telluride in 1913. For a period, he will be the local representative in Southern California for the Rand Venture Mines Company at Red Mountain in the Mohave Desert.

Holland was often a contributing writer to mining journals. One of his articles appeared in Mining and Scientific Press, October 19, 1918. Entitled "Recent Developments in Molybdenum" it concerns his work at the Climax Mine, Summit County, Colorado. World War One was still raging on when he wrote the article and the following brief quote gives a look at mining during the war years and after. "In common with most mines in the West. Climax is suffering from a shortage of skilled labor, and this is likely to become increasingly acute until such time as the civilized world has accomplished its most important duty of thoroughly beating the Hun. Meantime the American Metal Co.'s interest in the Climax Molybdenum Co. will be taken care of by A. Mitchell Palmer, as administrator of enemy alien property. The Climax Molybdenum Co. is fortunate in having an able and efficient technical staff. . ."

Mr. Holland was correct about the labor shortage which would continue after the war that ended on November 11, 1918 less than a month after the publication of his article. Lack of men would still be an issue for him later in Arizona as well. It would appear from the above quote that American Metal Company was an overseas company of an enemy nation and Holland wanted to be confident the readers understood he was not aiding the enemy in his mining efforts as Molybdenum was a strategic metal used for many purposes among them to harden steel. The rest of the article is an in-depth discussion of the type of geologic deposit the molybdenum ore is contained in and the methods and equipment used to extract and concentrate the recovered metal. It became instantly obvious that Mr. Holland was a very capable man. His knowledge of chemistry and geology were very evident in this article. The preparation of the Climax mine cost an estimated $600,000 and was a complicated tunnel system at nearly a 12,000-foot altitude often covered with heavy snow. Questions arise later about the qualifications and experience of both the superintendents at Meteor Crater during the 1920-1922 drilling operation. In the case of Mr. Holland, those criticisms can be disregarded; he has vast experience and knowledge.

Mr. Holland would be approached by T. A. Rickard the "owner" of Mining and Scientific Press to write an article about Meteor Crater. Holland will pass the proposal on to D. M. Barringer and not write the article himself seeing it as a conflict of interest and a breach of the no publicity rules of his employer in Boston. Mining and Scientific Press was published weekly by The Dewey Publishing Company of 420 Market Street, San Francisco. An annual subscription was just $3 for readers in the US and Mexico. Single copies could be obtained for only ten cents each. T. A. Rickard was listed as an Editorial Contributor on the masthead of the magazine in 1914 and directly under the magazine title appears "Established May 24, 1860 Controlled By T. A. Rickard." Like Holland, Rickard was a prolific writer on mining and mining engineering and apparently the man in charge at Mining and Scientific Press. L. F. S. Holland's name would often appear over the next half-century in mining journals, and it still appears in mineral resource databases presently.

Holland was sent to Meteor Crater, which at the time was a still-disputed geological structure in north-central Arizona. His task was to drill holes in the rim of the crater and locate a buried iron asteroid. Magnetic deflection studies had revealed a deviation from magnetic north interpreted to mean a large mass of nickel-iron lay buried deeply below the southern edge of the crater rim.

After a long struggle to once again get some financing, Daniel Moreau Barringer entered into an agreement with United States Smelting, Refining and Mining Company. Mr. Holland reported primarily to the office in Boston. Beginning in the spring of 1920 a subsidiary company named Crater Mining Company was created. L. F. S. Holland was their first superintendent at Meteor Crater.

The remains of the work on the south rim are still visible when you stand there today. Pieces of machinery are half exposed above the sand that has blown in over the nine decades since the workers left. A piece of pipe nearly even with the ground marks the location of the drill hole. Sections of water pipe are strewn around the site. Timbers lie about the area, and many more have been pushed over the edge of the crater.

Holland was the superintendent for the first 312 feet of the drilling. There were numerous problems with the tools. The drill rod was by several reporters not of good quality. Barringer called the materials "very inferior wartime products" which is a direct quote of what Holland wrote in his weekly report dated January 1, 1921. Barringer had no input about what was obtained for the work. Frankly, after his first years at the crater, he never again had enough money to do much work on his own. By the early 1920s, after two decades of work Barringer had spent over a $100,000 of his own funds at the crater. He had been successful in mining ventures in the Southwest. He could afford to spend money at the crater initially. Barringer quickly sought to finance the work through the sale of shares in various companies he created and through the assessing of the shareholders monthly in increasing amounts. He was wise enough to see that any large operation at Meteor Crater was too big for him and better done with other's money instead of his own.

As a Superintendent at an isolated location such as Meteor Crater, Holland was responsible for every aspect of the project. There was a drilling foreman who reported to Holland, and he reported to the home office weekly in letters or telegrams. He received telegrams and letters from not just his direct bosses but also from Barringer who was certainly keeping his hand in the operation.

Barringer further expected Holland to be the gracious host to any guest that he might send to the crater for a visit. Scientists, educators, personal friends and Barringer family members were to be greeted at the train station, taken on guided tours, and entertained with the story of the crater. But at the same time, the real nature of the work was to be concealed from some other guests to prevent the information becoming too public. Barringer did not want any differing opinions about the crater to appear in print. Also, since the time Arizona became a state, there had been a battle to keep the taxation of the property at a minimum assessment. Should the tax collector's office believe that millions of dollars of buried iron and nickel lay under the site, they would raise the taxes exorbitantly to help finance the fledgling state government. Holland was put in the position of having to walk this tight wire while handling all the other tasks both routine and extraordinary with perfect performance.

Barringer's communications with Mr. Holland are friendly. He also sends the best regards of his sons too. But in the years following the drilling, Barringer will be harsh in his criticism of both the superintendents of Crater Mining Company's Drill Hole Number 1. Barringer's use of the words "inexperienced superintendents" in his criticism will in the case of Mr. Holland be a grossly inaccurate characterization.

D. M. Barringer

The afternoon shadows are lengthening, and the Sun will soon fall behind the mountain to the west. The little boy sits on the veranda with the great man who occasionally speaks. In the woods around the house, the boy can see the men who hide behind the trees and then poke their heads out to gaze at their hero. After a moment they silently depart. They have both paid their respect and satisfied some need within themselves to see the beloved man. The man never speaks of these shadowy visitors who come every day. The boy does not understand all of the meaning of this. Yet he knows that the man is a friend of his father and that these weeks he visits him are important. General Robert E. Lee would say to the boy in one of those moments of conversation "My son always be a gentleman." Those who will later know the boy when he is grown to manhood will say of him that he was a gentleman.

He gazed out ahead of the train which was beginning to move. He was still standing on the step of the Pullman car as it headed out of the station at La Junta, Colorado. As he turned to look back and step fully into the car he saw her. She was running as hard as she could to catch the accelerating train. In that instant, before she would lose her battle against steam power she was close enough for him to reach out and pull her inside the car. La Junta is Spanish for "the meeting" and that was exactly what it was. But was it just that he was a gentleman helping a maiden in distress or was it also that he was single, traveling alone, and she was attractive. Or was it all those things mixed with some fate. Their journey on the train began a romance, leading to marriage and a life together. It further weaved him into activities in Arizona her home.

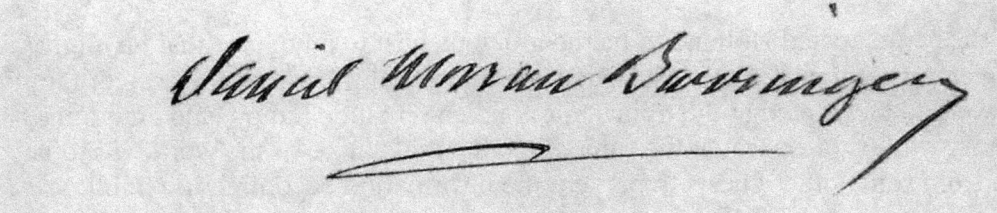

Daniel Moreau Barringer was born May 25, 1860, in Raleigh, North Carolina. His father was a member of Congress for three terms from 1842 through 1848 and a friend of Abraham Lincoln also a congressman at the time. After declining to run for a fourth term in Congress, his father was appointed by President Zackery Taylor to be Minister to Spain. It was into the rarified world of politics and power that Daniel Moreau Barringer the Second was born.

His mother died when he was seven and his father when he was only 13 years old. He was sent to Philadelphia to an older brother and in two years entered Princeton as a sophomore. He graduated a member of the famous Class of 1878 which included such notable individuals as inventor Cyrus McCormick, William Magie the physicist who will support his work later, and future US President Woodrow Wilson. After Princeton, he attended the Law School of the University of Pennsylvania graduating at the age of 22. He would briefly attend Harvard to study Geology and Mineralogy but transferred to the University of Virginia which was more challenging where he would also study metallurgy.

He joined the law firm of his brother, but his wild spirit was not then interested in the law. He loved too much adventure and big game hunting with his friend Theodore Roosevelt. Along with Roosevelt, Barringer and Owen Wister would help form the Boone and Crocket Club.

Eventually, he did find his calling in mining and geology. Working for a year at the Arkansas Geological Survey. Later with partners, he became wealthy from the Commonwealth Mine in Pearce, Arizona. He had both successful and unsuccessful ventures in mining throughout his lifetime. He settled down in Philadelphia with his wife to raise two daughters and six sons. All the boys would attend Princeton as he had. But he was never far in spirit from the West especially Arizona.

He was born into rarified society, and he would remain there most of his life. We call it networking today, and Barringer was a master of utilizing the connections he had with prominent personages in both government and academia.

His study of the law would become invaluable to him in later years. Even at the beginning of his career, he tapped into it by writing a masterwork entitled "The Law of Mines and Mining in the United States." The book was for a long time the authoritative work on the topic. But he would later use his legal training to defend himself in courtrooms as he attempted to develop Meteor Crater the real obsession of his life.

Such a long day at the mine and I am so tired. I must get something to eat before heading to the boarding house; I had no lunch. The mine is still profitable but slowing playing out. Two-hundred-fifty men work there now digging out 300 tons a day. It will stay open a few years, but someday I will have to rely on other ventures to support the family. The high-grade ore is mostly gone. One of the richest silver strikes in the country's history, and there has been plenty of gold as well. Maybe the crater will prove similarly lucrative. Millions of tons of iron and nickel and now they have confirmed the diamonds. We must beat the quicksand and reach the metal. The bigger shaft with the steam host is the answer I'm sure of it.

I must get some more cigars tomorrow before heading home in case they have none on the train. It is a long trip to Philadelphia, and I don't want to be without cigars. A little whiskey and a few minutes sitting on the porch will help me off to sleep. But I feel like a good dinner and some pie before that. I'll miss the food at the small cafe here. It's far better than the food on the train. Pearce, Arizona is not a big town but certainly much larger than before the mines opened. There was almost nothing here before that. Nine years since we opened the Commonwealth and now quite a little town has . . . Oh!

The meteor was so bright. I have never seen anything like it before. The light from it startled me. It is still afternoon, yet it was dazzling to my eyes. Blue white with jets shooting out at the front and the long fiery tail of yellow. The train of smoke persists in the sky. The bright pieces dropped from it like burning tar; then they fell away to the rear as the blue-white head continued on. They must be chunks of the large piece breaking off. How many were there, five? I think five pieces fell behind and slowly extinguished. Was that in a tiny way how it was when the crater formed? Are all the irons we've found around the hole pieces that fell off as it passed through the air? I must remember this date and write these thoughts down while they are fresh in my mind. April 11, 1904, the day I saw the meteor. I fear not even the glass of whiskey and relaxing on the porch in the cool evening air will be able to bring on sleep tonight. Not after seeing that meteor.

The meteor that Barringer saw on April 11, 1904, was a very significant event for him. He would report on it in one of his formal scientific papers about Meteor Crater. The meteor of April 1904 was seen by others in Arizona among them Samuel Holsinger who first told Barringer of iron found around a great hole two years prior. Barringer would also find a small stony meteorite on the ground some distance from the crater. He would wonder about whether it was a separate fall or part of the main event that formed the crater. Thoughts of a "raisins in bread" composition of the asteroid would plague him for a time. He wondered about if the asteroid was solid metal or a mixture of stony material and iron. He gradually changes his belief that the crater formed from a single large mass, believing later in a swarm of smaller masses. He will sketch a buried swarm on cross-section drawings of the south rim but use wording after the drilling program to say they have without a doubt found the main mass.

It is a little hard a hundred years later to grasp the level of ignorance about asteroid impacts that existed during Barringer's lifetime. No prior impact crater had ever been discovered. Nothing was known about the processes involved. It was only being suggested for the first time that the craters on the Moon might not be volcanic, but instead formed by objects hitting our neighbor. After dozens of movies and TV science fiction programs on asteroid collisions, it is common knowledge that some of the craters on Earth are from impacts. The probes we have sent into our solar system, reveal most bodies are heavily scarred from brutal poundings that never disappear through erosion or tectonics.

It was into this void of knowledge that Barringer stepped. For nearly thirty years he sought to pull others in as well. Always the lure of vast amounts of surviving nickel-iron as the bait used to find financing. It is interesting that with all the many persons that lost money in the search for buried wealth at Meteor Crater none of the shareholders of any of the companies seemed to be angry about the failures. They seem instead to have caught Barringer's excitement for the project and the discovery of something new and cosmic in nature. The only person to be outspokenly discouraged was Barringer's first partner Benjamin Chew Tilghman.

The equivalent of millions of dollars today was spent at Meteor Crater searching for the buried asteroid. The drilling program on the south rim begun by Mr. Holland was just one of the attempts. It would end up costing approximately $200,000. The results would be misinterpreted as positive and spawn another even more expensive attempt to reach with a mine shaft what they believed the drill had found.

What is an obsession? The clinical definition is "the inability of a person to stop thinking about a particular topic or feeling a certain emotion without a high amount of anxiety." Was D. M. Barringer obsessed on the topic of buried wealth at Meteor Crater? There is no doubt based on the frequency of letters, telegrams, and meetings over a span of nearly thirty years that the crater was never far from Barringer's thoughts. But there is no evidence that he suffered anxiety over the crater until near the end of his life. That was the period when the reports of Forest Ray Moulton made it clear that the asteroid did not survive the impact. At this point, Barringer was emotionally low and unhappy. He tried to rally his supporters after Moulton's first report. Later more refined and detailed reports made it clear that the iron asteroid mostly vaporized on impact. Barringer was troubled by this news. It is never easy to lose one's dreams, but was he obsessed?

The correspondence that Barringer has with Holland during the drilling program on the south rim is genial and level-headed. He appears to be a fully functioning individual with a family life and other business concerns. He does not seem obsessed, passionate about his belief that there is great buried wealth at Meteor Crater, yes. As an American businessman in the mining industry, he wants to see that resource exploited.

There was nothing of con man about Barringer either. He had a vision of great wealth. He needed funds to reach his goals. He needed amounts of money which he did not have himself. His ventures were seen with mixed feelings by some, and at one point he needed to prove that he had not promised more than was legal. Leading other dreamers was not criminal then, and it is not illegal now as long as your followers know the facts and if all the risks are disclosed. That is pretty much the way investments have always been. You put your money out there with the hope that the economy will be vital and on the rise and that you will make money. If the economy is sluggish or declining to depression, then you may lose your investment. As it turned out the court in Pennsylvania decided that though Meteor Crater was a very speculative investment, Barringer was doing nothing illegal in attempting to raise money by selling stock. He was allowed to do so.

Neither Barringer or his investors in any of the businesses he formed to exploit Meteor Crater ever made money. In fact, the investors of the original Standard Iron Company were assessed ever-increasing amounts as stockholders. Long after his death, the silica mining on the south slope may have brought in a little money. But they moved only a few hundred tons of silica from a reserve estimated to be more than a million tons. It too was only a small operation, short-lived, and the silica was priced very low.

Mining is a difficult industry. Every day that you are in operation, you are closer to the day when your resource will be depleted, and you will be out of business. With the lumber industry, you can create a sustainable approach and replant trees. With mining once the ore is extracted and the veins run out that location is finished. Barringer was in that situation. His great early successes were providing a diminishing return. The Commonwealth Mine, in Pearce, Arizona which had been so profitable initially was nearly finished by when the drilling on the south rim of Meteor Crater was undertaken by United States Smelting Refining and Mining Co. Barringer and his partners had sold the Commonwealth in 1910. It had been a big operation for part of the time. They had an 80 stamp mill by 1901 which replaced the 60 stamp mill built in 1898 and which burned down in 1900. In the early days, the metal bars were made so heavy that they could not be carried on horseback because of the danger of robbery. The gold and silver were moved by wagon train to Cochise until the Arizona and Colorado Railroad built a line from Cochise to Pearce in 1903. During those years the former Deputy Sheriff of Pearce a Burton Alvord had gone from lawman to outlaw and joined with another badman, Billy Stiles. The Alvord-Stiles gang robbed trains and committed other crimes using Pearce as their headquarters. The population during the boom years had swollen to 1500 souls, and the city had everything including a movie theater. The mine is still the subject of interest today. Some of the tailings have been reworked and other waste rock shipped away for retreatment. Modern tests indicate that 2.5 ounces of silver per ton is in some of the waste rock from the early days. With about one million tons of tailing and dump rock on the site, there is an incentive to revive the recovery of the precious metals again. Production was intermittent at the Commonwealth for decades after Barringer's time. Nearly 20 miles of tunnels are underground.

Today Meteor Crater is still a natural wonder. There are scars from some of the work easily seen by visitors. The machinery is mostly gone, and it is a tourist attraction. It does not take much sometimes to catch a lifelong passion. Sometimes it is as easy as seeing a brilliant meteor streak across the sky that turns you from one road onto a completely different path and creates a passion. And Barringer was a man who indulged his passions. He hunted and rode horses, loved a good cigar, and he drank whiskey. Prohibition was in force during the years of the drilling on the south rim and Holland never speaks of any alcohol at the crater. But the 18th amendment did not concern the drinking or private ownership of alcoholic beverages only their manufacture and transportation. Barringer had as his son Brandon would write several decades later, put aside for himself before prohibition went into effect a supply of whiskey "to last a liberal expectation of a lifetime."

Barringer had not put everything into his dreams and died pennilessly. But he sold the family home and they lived toward the end of his life in a series of rentals. He had some business activities besides the crater and all his six sons went to Princeton. But at the time of his death he was no longer the wealthy man who had started the work at Meteor Crater in 1903. The family is still associated with the crater property. They have abandoned his search for the asteroid and have shown a commitment to preserving the site as a natural treasure and a research location.

With the growing understanding of cosmic impacts, even D. M. Barringer might have given up the search at some point had he lived. He was passionate but also pragmatic. After the final shaft failed, even he might have given up the quest, but he died before the end of that endeavor out on the south slope. Approximately $600,000 was spent at the crater to find a great mass of iron that was likely never there. The best evidence for it vaporizing was in front of everyone from the beginning, tiny metal spheres. What trillions of them in the soil around the crater meant was not revealed until H. H. Nininger came to the crater almost two decades after D. M. Barringer's death. The asteroid had vaporized, as the vapor cooled it turned back to solid metal. A rain of tiny metal droplets followed. Some of them rusted away over 50,000 years, but others with more nickel and cobalt survived.

Barringer may never have found the vast quantity of nickel-iron asteroid that he searched for, but his work has left a great legacy of knowledge none the less. The many iron obstructions hit by drill bits during his explorations tell us that there are iron meteorites buried under the crater floor. The shafts he dug have provided scientists the information on the stratigraphy of the crater bottom. Access underground is still available even a hundred years later by one of the Barringer era shafts. Meteor Crater has fired the imagination of many since Barringer, and that certainly includes one boy in the 1960s who visited for the first of many times and found himself a little obsessed.

William Francis Magie

I reached out and then I saw it and hesitated. I want to help, and I am intrigued with the physics. And he is a Princeton classmate and friend. But his letters come more frequently, and his problems are taking much more of my time. After the hesitation, I take the letters from the mailbox. His was on the top, so I saw it. I guess if it had been mixed in I would not have found it until I got to my office. As it is I should not read it now, I have a class for undergraduates next hour. I have one student coming for a thesis conference. This letter from Daniel will get opened this afternoon.

I have to admit that I enjoyed my visit to his crater. Holland, the superintendent, was very accommodating and informative, he gave us free access to everywhere. And the crater itself is a wonder. I am convinced it was made by an asteroid plunging through the atmosphere. But I am not sure if I agree with Daniel that there is a huge amount left to be found. I doubt that it hit at just 2 miles per second. Bodies moving in space are going much faster than that, and the atmosphere should not have retarded an asteroid big enough to blast out such a hole. Something more like 8 miles a second for an impact speed. I'm sure that is what this letter concerns. Either that or he wants to know what I think about the magnetic survey.

Should I just open the letter and find out while I am still walking. If I do, it will be over, and I can get on with my day. I will have to answer it regardless I might as well know. It is just a copy of a letter sent to all the group of us. It is about the magnetic survey and about testing some of his shaleball meteorites while they are in the ground to see if they have a magnetic polarity different from the local magnetic north indication. I have to write them back especially Mr. Moore in Boston. I did that study myself when I was there. It would be an enormous waste of Mr. Holland or Mr. Fay's time for them to repeat the work. The measurements were so varied and confused on each specimen that I realized there was nothing that could be determined and applied to the crater as a whole. Today is Thursday if I get the letter in the mail, Moore could have it on Monday the 20th and send a telegram to Holland.

Daniel had a good idea though one I tried myself. I wish he had been at the crater when Mary and I were there. He could have seen the experiment. It is too bad the meteorites do not have just a single magnetic polarity. I have a feeling that this debate about the crater is going to get much more intense before it is resolved finally by science rather than opinion.

For nearly fifty years Professor William Francis Magie taught physics at Princeton University. He was Dean of the Faculty from 1912-1925 and Chairman of the Physics Department for much longer. He was born December 4, 1850, in Elizabeth, New Jersey. He was an excellent student graduating valedictorian of his class. But he had no real direction until offered the position of assistant to Professor Brackett on Commencement Day. He took the offer, and his introduction to science turned into a lifelong love of physics. After writing a research paper that was well received he took a leave. He used the time to study in Germany where he received his Ph.D. from Berlin University.

Magie returned to Princeton and taught for the remainder of his life. He wrote several books on physics and did some research that was revolutionary for its time. He co-authored the first paper in America on the use of the newly discovered X-rays in surgery. He retired in 1929. He married Mary Blanchard-Hodge on June 7, 1894. Magie died on June 6, 1943, and is buried in the Princeton Cemetary.

During the years before and after the time of Mr. Holland and the drilling at Meteor Crater, Professor Magie was one of D. M. Barringer's most trusted sources for scientific answers about the crater. During the fall of 1920 Magie and his wife visited the crater. Mr. Holland was their host. He had been instructed to give them free access and to make their stay enjoyable. During Magie's visit, he did a study with a compass of the magnetic fields of shaleball meteorites. Trenches were dug on the slopes of the crater to expose the meteorites. He made his measurements where the shaleballs lay without moving them. His results though inconclusive were a very early use of an instrument to read the recorded magnetism of rocks. A hundred years later this is the established area of geology called paleomagnetism. Rocks are tested with modern sensitive tools to read the polarity of the magnetic field of the Earth as recorded at the time the rocks crystallized.

The period that Holland was at the crater was a relatively quiet time for Magie. As the debate over Meteor Crater in the scientific community would not heat up until the late 1920s. Barringer would then lean on Magie and others for more information and opinions.

Magie was dissatisfied with the results of the extensive magnetic survey of the crater done in 1920. The results were vague, and possible anomalies were not further surveyed to define their extent. The same view all the parties involved came to after seeing the charts of the survey.

Some individuals abandon a friend in a heated debate. Magie never abandoned Barringer. Characterized by biographers as a man with a kind and caring heart Magie was also known as a man that did not yield under pressure in a debate. He could remain a gentleman while holding to his view. He must have had a few discussions with himself as the evidence began to turn against Barringer's beliefs about the crater. But the physics was what he knew, and relating the physics to Barringer though hard was what he did.

Magie retired from Princeton in 1929 just as the mathematical reports about the crater were compiled by Forrest Ray Moulton. Barringer would die in late November of 1929. Many think that it was anguish over the proof in the mathematics that the asteroid vaporized which lead to Barringer's heart attack. So, in the end, it may have been the physics that Magie loved which contributed to his friend's death. Magie is retiring to years of enjoyment and family life while his classmate is passing away. Both men have achieved a place in history. Magie is always a character in the Barringer story. Barringer, however, is not mentioned in biographies of Magie. They began together at Princeton but ended in very different places while remaining in contact. Not an uncommon fate for friends.

Elihu Thomson

The cool breeze feels nice on my face. The sky is clear and the evening is still. A perfect night ahead I think to sit in the observatory and gaze upon some distant islands out in space. I remember so many nights when I have looked up at the stars and many nights when my eye barely left the eyepiece. I have loved astronomy since that night when I was five and Donati's Comet was hanging in the sky like a giant scimitar. If I close my eyes now, I can see it clearly in my mind. Nothing had such an effect on me as a youth like the great comet of 1858, except for maybe the meteor showers of 1867. Over half a century later and I remember the bright streaks of light coming ever few seconds and the moments when several meteors crossed the sky at the same time. I pondered then what the objects were and how fast they would have to travel to become incandescent and burn completely up. Maybe that is why I am so interested in the work that Barringer is doing out in Arizona at his giant hole. I am convinced by what he has found that the crater formed by the fall of a great asteroid from space. It is the location

of what remains that is the current mystery. Standing there on the rim of the hole was another one of those life events that I'll never forget. Like seeing the greatest comet of the 19th century, I will always remember looking across the open chasm toward the rock walls on the other side and imagining what being there at the time it formed was like. I still can not push my imagination far enough to conceive a conflagration sufficiently intense. Barringer's envelope that came today had a letter from Magie and the magnetic deflection chart of Mr. Fay. My life has been a study of electricity and magnetism. There is something there buried under the south rim. Magie agrees that there is something near Station No. 8 too. If they had a dip needle instead of just the compass and did more readings further off on the slope, the magnetic inclination at those places would point to where the mass is buried. The magnetic survey must be studied with great care, but I wish it were more detailed.

Elihu Thomson is sixty-seven years old when D. M. Barringer is seeking his wisdom about the magnetic survey done at Meteor Crater. Thomson was certainly one of the foremost inventors and innovators in the field of electricity. He will hold 700 patents and receive many awards during the course of his life. Though mainly known for his work in developing lighting and welding his contributions cover many areas of science and industry beyond electricity. Thomson was the first to suggest that a mixture of helium and oxygen could be used by workers in pressured caissons to prevent the bends. This is, of course, a gas mixture that is used by deep divers today. His life-long love of astronomy led him to expend significant effort to make fused quartz glass for use in the optics of telescopes. The available glass of the time was unsuitable for large mirrors. It was full of stress and changed with temperature in an uncontrollable way distorting the astronomical images. The technology of the time did not allow him to make large blanks of fused quartz and pyrex glass was created around the same time. Pyrex was chosen for the great telescopes like the 200 inch at Palomar. But fused quartz is used in many applications today including telescope optics.

Elihu Thomson was born on March 29, 1853, in Manchester, England. The family immigrated to America in 1858. The same year that he saw Donati's Comet. An image of which stayed with him all his life and began his love of astronomy. He was a bright student. He taught at Central High School in Philadelphia until along with a fellow "professor" designed an arc lighting system and formed the American Electric Company. The name changed in 1883 to the Thomson-Huston Electric Company. In 1892 his company merged with Edison General Electric Company.

In 1899 he completed his home observatory having made all the parts himself including grinding, polishing and figuring the 10-inch mirror. He later wrote a book describing the process of creating an astronomical telescope mirror.

Thomson was President of MIT from 1920-1923 and a lifetime member of the corporation as well as a lecturer there. During the time when he was President of MIT, Crater Mining Company was drilling for the buried asteroid at Meteor Crater.

Barringer had developed a friendship with Elihu Thomson over many years. There had been meetings with C. W. Moore and Jennings of United States Smelting Refining and Mining which included Drs. Magie and Thomson. Thomson wrote two articles on the crater in 1912 after his visit there in April of 1911. In both the papers, he expressed his total belief that the crater had formed by the impact of an asteroid. As early as these 1912 papers he wrote that the mass had come to rest under the south or southwestern portion of the crater wall. He was one of the proponents of the slow impact speed believing that the asteroid had hit the ground at just 2 or 3 miles per second.

Barringer was at times frantic in his attempts to get financing. He literally asked everyone he knew. From former President Theodore Roosevelt to even Elihu Thomson. No one escaped his pleas for help. On two occasions Barringer tried to get Thomson to invest in the drilling on the south rim. Barringer offered to create a lease with General Electric in which Thomson was a major participant. Thomson was not persuaded. Later Barringer wanted Thomson to arrange for MIT to drill at the crater. Thomson was president of the corporation at the time. Again Thomson would not invest or become involved. Professor Magie likewise received pressure from Barringer. Barringer tried to get Princeton to undertake the operation at the crater.

Elihu Thomson aided by Dr. Magie was able to change Sidney Jennings' mind on the nature of the impact and the probability of the asteroid remaining in one piece. C. W. Moore of United States Smelting Refining and Mining had been convinced earlier, but Jennings believed the asteroid had been disrupted into fragments that were spread everywhere under the crater. After meeting with Thomson and Magie and hearing of possible competition having an interest in the crater he was ready to approve the drilling. Jennings arranged to meet with Barringer to finalize the lease in early 1920.

While the site for the drilling was prepared magnetic surveys were conducted, and Thomson along with Magie was very involved. Both made suggestions about the methods the survey should employ. Both will later study the results. Elihu Thomson sent sketches and recommendations to Barringer that were passed on to Mr. Holland. His sketches showed how the indications of buried iron would appear in the readings obtained. He drew in his sketch how the magnetic lines of force would bend around a buried iron object. Later he would along with Magie be very disappointed that more readings were not obtained to define the location and size of anomalies appearing in the survey.

Thomson would remain a staunch supporter of Barringer's belief that the crater formed by an asteroid impact. He would be called upon more and more as the debate grew within the scientific community to offer wisdom and express his opinion and share his knowledge. But as with Magie, his life was much larger than Meteor Crater or Barringer's needs. Thomson had a wife and a family life and his work as an inventor. MIT and his 11 children filled his life as did his home observatory with its handmade 10-inch telescope. Elihu Thomson lived to the age of 84 years and lived passed

Barringer as Magie did. Thomson died March 13, 1937, leaving a legacy of invention that is continuing to benefit humanity today.

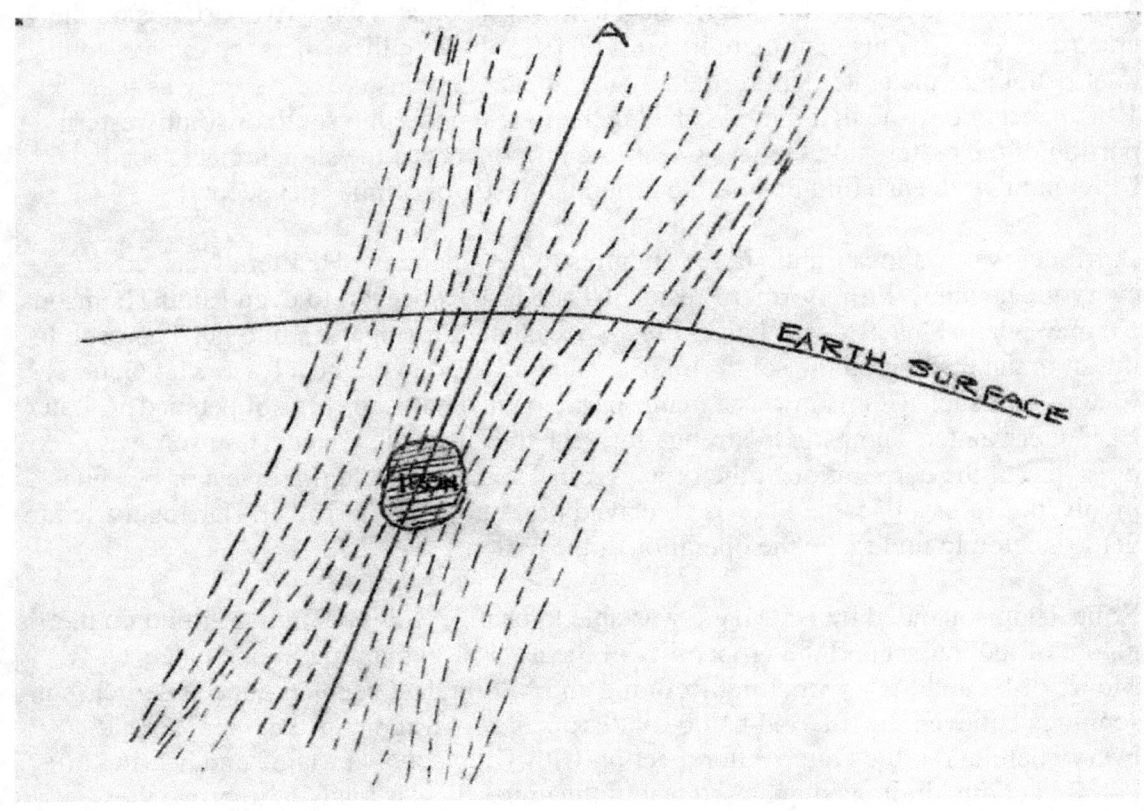

Sidney J. Jennings

"I will not be in the office tomorrow, please change all my appointments until the afternoon of the next day or see if Mr. Anderson is willing to take my meetings. I will be catching the early train, and it is a long day to Princeton and back. I may be in late the next day."

Well, that is the last of my business for today. I have to get home and get ready to leave early. I have thought for months that this business with the crater was over. But now with Ricketts and the Arizona Copper Boys showing interest in the property I have to give the project another look. I was out there with Moore, and we studied the site for three days. There is iron with some nickel, and I am personally convinced Meteor Crater formed by the impact of an asteroid long ago. But I am still very doubtful that it is in one mass buried somewhere under the crater. Everything I saw tells me it was shattered into bits that are scattered.

I am open to hearing what they have to tell me. No man in America has more qualifications than Dr. Magie to discuss the physics of the crater and methods to find the asteroid beneath it. If nothing else this may prove to be a unique mining operation. If we go ahead with it.

The car will pick me up here to take me to the train; the campus is almost at the train station in Princeton, so I can walk from there. I have the office number, but it should not be hard to get someone to direct me to Magie's office. He is the Dean of the Faculty from what my research assistant said. It would almost be worth spending the money just to hold off Ricketts from moving into iron; his group already has most of the Arizona copper. Barringer is saying 10 million tons, with platinum and iridium. Others are saying much less could make the same size hole. I'll see what Magie says. They have me talking to Thomson also. I feel like my arm is being twisted a little, but it is a pleasure to meet these two famous men.

I have not been on a campus since my own graduation from Harvard. This trip may be fun, but it will be a long day, I have to get some sleep if I am going to meet the early train.

Eighteen years in South Africa working as consulting engineer and manager for some of the most significant gold and diamond mines in the world made Sidney Johnston Jennings a man of high regard in the mining industry.

His early education had been in France and Germany. He graduated from the Lawrence Scientific School of Harvard University in 1885 with a degree of C.E. (civil engineer). He spent just a few years working here in America before heading to South Africa in 1889. He traveled with his older brother J. Hennen Jennings who also would have a distinguished career in South African mining. Sidney's first position was as Manager of the Willows Copper Argentiferous Syndicate. Soon, however, he was appointed Assistant Manager of the De Beers Consolidated Mines. He worked in the diamond industry until 1893 when he left Kimberley to be the Manager of the Crown Deep Ltd. one of the subsidiaries of the Rand Mines of Johannesburg. There he established a record for the fastest driving of a tunnel through rock. His record remained unbroken for many years. September 1, 1896, he was appointed to the position of General Manager of the Crown Reef Gold Mining Company where he stayed for three years. He more than doubled profits by introducing a process to sort the waste rock, and by creating a contract system for paying labor.

Jennings was appointed the Consulting Engineer of Robertson Gold Mining Company in 1899. They were the largest gold producer in South Africa. He was also appointed Assistant Consulting Engineer to Messrs. H. Eckstein & Company the largest mining firm in South Africa.

The Anglo-Boer War erupted, and Jennings went to England. He returned to South Africa in time to be among the first group of consulting engineers allowed back into Johannesburg. Louis Seymour who had been Consulting Engineer for Messrs. H. Eckstein & Company and Jennings superior was killed during the war, and his position was filled by Jennings.

Sidney Jennings was appointed to the town council of Johannesburg by Lord Milner and was elected Chairman of the Works Committee. He was a corresponding member of the Council of Mining and Metallurgical Institute of London. Jennings was Vice President of the South Africa Association for the Advancement of Science. He will later be the president and a life member of the American Institute of Mining Engineers.

In 1907 Jennings returned to America and established himself in New York City by starting a company of consulting engineers. He was the Consulting Engineer for Boston Consolidated until they were absorbed by Utah Copper Company. In 1908 he joined United States Smelting Refining and Mining Company.

By the time he sends Mr. Holland out to Arizona, he has been with United States Smelting Refining and Mining for years as the Vice President of Explorations. The company has by this time vast holdings and is reporting millions of dollars of bottom line profit. Jennings as the person in charge of explorations was approached along with C. W. Moore, Chief of Engineers by D. M. Barringer to take a lease at Meteor Crater. That had been back in 1918. The two officers had visited the crater for three days in the last week of April. They collected specimens of the iron meteorites and iron shale for analysis. But they ultimately do not accept Barringer's offer. Jennings seems from reports to have been the most reluctant to invest at Meteor Crater. He was convinced that the asteroid was pulverized into thousand of fragments now scattered under the crater floor. He was aware of Barringer's troubles with quicksand at around two hundred feet down. He knew from Barringer's scientific articles that the fragments of meteorites the drill had struck on the crater floor were down far deeper than the quicksand. Mining the floor was going to be difficult or impossible.

After United States Smelting Refining and Mining turned down the lease Barringer sought to finance the drilling everywhere he could imagine. This frantic pursuit for money brought Barringer into contact with Louis Davidson Ricketts associated with the massive copper mining industry in Arizona. Word of Ricketts possible interest in Meteor Crater got back to Jennings and revived the discussions about a lease for the exploration and recovery of the iron. Jennings still needed some convincing that the asteroid had survived intact. A meeting with Dr. Magie was arranged at Princeton. Magie a classmate, friend, and supporter of Barringer on the cosmic origin of the crater was the perfect person to discuss the matter with Jennings. Magie was an acclaimed Professor of Physics and Meteor Crater is if nothing else a problem in physics. Jennings also met with Elihu Thomson for a similar discussion and afterward approved the idea of drilling at the crater. They planned to drill down from the south rim. Barringer a long believer that the crater was a confined pool of water, believed that the ground was dry around and beyond the hole itself. The lease was signed, and almost immediately Mr. Holland is sent to the crater to prepare the drill site, run a pipeline to supply the rig and camp with water and oversee the operation of the drilling after the rig was erected.

Jennings will visit the crater once during Mr. Holland's time there. His visit in early December 1920 was at the very moment when the first serious problems were occurring. His vast experience in mining from his years in South Africa seems to show. He is not noticeably bothered by the difficulties. He considers them just a normal part of drilling in fractured rock. He leaves with his mind more on getting an inventory of the surplus equipment and supplies then on the blasting off of the casing shoe and the bottom fourteen feet of casing.

Sidney Jennings will receive weekly reports from Mr. Holland for most of the year that Holland is at Meteor Crater. At first, Holland is sending his reports to others. Soon the reports are going to Jennings with copies to other men. Holland is receiving letters and telegrams from Jennings nearly weekly in response to his reports. Some weeks there is little progress made. This was especially true as the drill site was being prepared. Lack of workers and other issues stalled the work so that Mr. Holland had nothing new to report some weeks. Once the drill encountered problems, Holland's reports began to read as a weekly report of misery. Jennings seems to maintain an even keel during all this; never expressing any dissatisfaction with Holland. But he is well aware that the agreed upon money has been spent and that they are bleeding money down the drill hole from December 1920 onward.

In February of 1921 when the underreamer bit is proved to be impossible to drill through, recover or pass, Jennings is thrown into a whirlwind of telegrams. Moore is telegramming Jennings; both are telegramming Holland, Holland is replying, they are replying back. Do they move the drill rig a little and drill down again, do they move the rig 20 feet and drill down again, do they abandon that area of the crater rim and move to another area that showed a strong magnetic anomaly, reerecting the rig there? Or do they dig a tunnel to clear away the broken underreamer tools and continue drilling down where the rig is? Holland is receiving instructions to wait at first, then to hire the surveyor back and locate where to drive a tunnel. Then a telegram saying do not start digging the tunnel until confirmed by others. Finally, after about two weeks of telegrams, Jennings tells Holland to dig the tunnel. Holland is to drive it into the crater wall at least 20 feet below the point where the broken bit and bottom of the drill hole are.

The tunnel digging begins, and Holland reports to Jennings the progress. At first, all is well then problems again become the topic of his weekly reports. Cavings in the tunnel, men with bad lungs from the dust, runs of sand that continued for days and threatened to swamp the tunnel are the regular stuff of Holland's reports.

CABLE ADDRESS,"SMELTINGCO."

UNITED STATES SMELTING REFINING & MINING COMPANY
55 CONGRESS STREET
BOSTON, MASS.

Number 822 Kearns Building,
Salt Lake City, Utah,
7 April, 1921.

Mr. L. F. S. Holland, Manager,
 Crater Mining Co.,
 Crater, Arizona.

My dear Mr. Holland:

    We have recently arrived at the decision to shut down some of our mining properties. This leaves a number of employes to be taken care of who have been in our service for a number of years.

    Owing to the fact that you have but recently joined our forces, it seemed fair to us to ask you to resign in favor of one of our other officials who has been in our service a much greater length of time. I would, therefor, like it if you would resign as from May 1, 1921.

    I will give the gentleman, who will be appointed to take your place, a letter of introduction to you and I trust that you will see he is properly accredited at the bank and introduced to the various people with whom we do business in your section of the country.

    I regret the necessity of the severance of our relationship, which has always been pleasant, and I will be very glad to assist you in any way possible to obtain employment.

I beg to remain,

Yours truly,

Sidney J. Jennings

In April of 1921, Jennings sends a letter to Holland requesting that he resign as superintendent effective May 1, 1921. Jennings says that the company is cutting back on their explorations and closing some properties. To keep employees who have been with the company longer Holland is to be replaced. Holland immediately replies with a letter of resignation. Jennings offers to give Holland a reference and assistance in finding new work. On May first Mr. Holland is unemployed and Mr. Plumb is superintendent.

The tunnel is only half dug when Holland leaves, but it is completed, and the hole is cleared. When the drilling begins again it is reported by Barringer's son in a biography of his father that the drill is now a rotary drill. It had been a churn drill during Holland's time. If this is the case, then Jennings and United States Smelting Refining and Mining have made another investment in equipment at Meteor Crater. They have changed the rig and the tools. By the time that the drill hole finally reaches its deepest depth of 1376 feet and is irredeemably lost nearly $200,000 was spent. There is no other way to view the south rim drill program than as a disaster. They learned nothing substantial from the drilling. The asteroid was never discovered. Two masses of iron meteorite may have been drilled through, and an area of iron shale nodules was certainly drilled through. But nothing was found to warrant further mining by USSR&M Co. Their chemical analysis revealed that much of the nickel content had leached away out of the iron shale. They would give up the lease in about a year ending their relationship with the crater. Barringer would run with the little evidence he thought had been found. He would lead others into a much larger financial disaster with the final shaft on the south slope.

Jennings grew up around mining. His father owned a coal mine in Hawesville, Kentucky where Sidney was born in 1863. Years later as part of his work with USSR&M Co, he would establish the mining town of Hiawatha in Utah's rich coal region. Jennings provided social welfare services for the miners, and it was considered to be a model community. Sidney Johnston Jennings died November 17, 1928, in New York City.

The Foreman Mr. Wammock

I wish that his first travel job was not to this isolated camp. He is my boy, and he wants to follow in my footsteps as an oilman. But he is only fifteen, and this is going to be very different than working near home in Fillmore. He has always had problems. We knew that right off when he was little. He did not catch on to things fast. His teachers called him slow, and it always made me start to weep. I would push the tears down. I am a big tough working man, and we don't show that stuff. But as it is doing now it broke my heart just thinking about it. He struggles with things that everyone else takes for granted. He is a sweet boy with a good heart, and we have tried to shield him all his life. But now, out here, he may have to learn to take what comes. The men he will work with won't understand. I have to pray they will not be cruel.

L. F. S. Holland was the superintendent, and he was responsible for the operations of Crater Mining Company. But it is evident from his documents that he depended upon the drilling foreman for the day to day decisions at the drilling rig. When Holland is asked to resign by Sidney Jennings, he contacts his other boss to ask for an additional letter of reference beyond the one Jennings offered. In this request to Mr. Anderson, he expresses his hope that he might be employed by USSR&M Co. again in the future. He states that there are few individuals in the mining industry with as much experience in mine examination as himself. But mining engineering and investigation are different than drilling. Holland lived in Hollywood, and much of the Los Angeles area was a forest of oil wells in the 1920's. Holland undoubtedly had some knowledge of drilling since it was common to drill and core in mine areas searching for ore veins. But he makes statements several times in his correspondence that indicate he depends on the experience and cleverness of the drilling foreman to overcome problems and to guide the actual drilling.

There were two drilling foremen during Holland's year at Meteor Crater. The first was Mr. Mort Branson who Holland describes as "a dead one." It is clear that he was ill when he arrived at the crater and that the altitude was further complicating his poor health. He would be stricken in the night September 13, 1920, with paralysis and have to be evacuated by train to Los Angeles. This happened just before the arrival of the drilling crew Branson had selected.

Branson was replaced by Mr. Wammock. W. E. Coan of the Oak Ridge Oil Company of Fillmore, California recommended both men. Mr. Coan's letterhead has Fillmore printed on it as the city, but most records show Santa Paula next door was the later location of the offices of Oak Ridge Oil. Mr. Coan was personally known to Mr. Holland who respected his opinion about drillers. It strongly appears from the documents that the selection of the foreman was based solely on Mr. Coan's recommendation. Likewise, the selection of the actual men in the drilling crews was based on the knowledge and experience of the two foremen. Mr. Branson had selected a crew he was familiar with, and they never came because of his illness. After a delay,

Mr. Wammock selected a group of men for his crew. Holland's only involvement was to accept the men and put them on the payroll.

Where Branson had been "a dead one" in Holland's opinion, Mr. Wammock was described as a strong "huskier" man. Mr. Branson was to be just the foremen and do no actual labor on the drill rig. Holland negotiated with Mr. Wammock to work one of the shifts as part of the crew for $15 per day. Mr. Wammock was likely the highest paid man at the crater. Still, Holland was saving the cost of an additional driller by having Wammock work one shift.

On January 26, 1921, Mr. Holland wrote a letter to Mr. Coan about the many difficulties they were experiencing. Mr. Coan in his reply states that he has "confidence in Mr. Wammock's ability to handle this or any other drilling proposition. I have known him a good many years and know he is a driller of wide experience." Mr. Coan will go on to write "From a perusal of your list of troubles, it appears to me that the crooked holes, broken tools, etc. are due to fissured limestone formation and natural conditions rather than to the lack of knowledge or ability on the part of the foreman." While Mr. Coan's comment is mostly correct from a historical perspective, it is possible that Mr. Coan was so supportive of Mr. Wammock partly because he had recommended him for the job.

The difficulties of the drilling were so severe at times that it is easy to see why Mr. Holland would have questioned the qualifications of Mr. Wammock. The Barringers placed the blame for the failure of the drill to reach its planned depth on the drillers. But as Mr. Coan has noted the blame was mostly the poor quality of the tools and the characteristics of the rocks they were drilling. In point of fact, foreman Branson in August 1920 upon seeing the drill stems provided for him to use tells Holland "they will not stand up." Holland has already reported that he feels they are too light himself and has ordered stronger ones like those used in California. Unfortunately, the men get little opportunity to use the heavier ones because they are slow in arriving. And the lighter ones which continuously break take such a long time for repair that Holland makes a trip to Los Angeles. He was forced to find out why so much time was being lost. Getting no acceptable answer from National Supply Company in Torrance he arranges for another company to supply and repair the drill stems. The Regan Forge in San Pedro, California was his choice. They were just a handful of miles away from National Supply and agreed to make the crater's needs a shop priority.

Mr. Holland tries to give as much consideration as possible to Mr. Wammock and to extend to him the benefit of the doubt about the causes of all the difficulties. But by February 1921, conditions are quite desperate. The underreamer lost early on has managed to find its way under the bottom of the casing. It is as Holland says "reposing" there completely blocking the hole. In his letter to Mr. Coan, Holland comes right out and asks "Is Wammock's judgment safe to follow?" He adds that he likes Wammock and if he has done as good as any foreman could have he would like to give him credit for it. But if Wammock "is not the man for this particular job, the sooner we make a change the better."

Within days of Mr. Holland letter in early February 1921, the drilling efforts were stopped when Wammock declared the hole lost. The next months were occupied in digging the tunnel to remove the underreamer and other debris from the bottom of the hole. Mr. Holland was asked to resign while the tunnel is still incomplete. Beyond a vague statement that may indicate the foreman was still at the crater during the digging of the tunnel, nothing further is recorded of Mr. Wammock.

Mr. Wammock had brought his son to work as a tool dresser. This presented a few problems and created some bad feeling among others on the crew. He was apparently a young man for Holland uses the word "boy" to describe him. But he had to be at least a young adult and would have needed to be a strong individual to perform that job. The bits he would reshape and sharpen weighed hundreds of pounds. They were not moved easily even when wheeled carts were available. Using the forge and sledgehammer was also not work for an actual boy. Another source reveals that he was fifteen years old. But Wammock's son had difficulties not received very well by the drill crew. Holland writes in a later letter to Mr. Coan on March 11, 1921, that "The boy is of small intelligence and has some filthy habits. As an example, he ruined three of the Company's new mattresses and had his room smelling like a stable. Some of the men suggested that he was too lazy to get out of bed to urinate. These things all come in the day's work and we must grin and bear them. . ." Mr. Holland would never have written this to his bosses in Boston. To have done so would certainly have raised questions about his ability to select or keep proper employees. But he felt free to discuss these things with Mr. Coan who he knew beyond this job and who had recommended Mr. Wammock. There was no excess of compassion for the young man by the rest of the crew but there may have been a little from Mr. Holland who was at least willing to accept the problems and move forward day by day. It was Arizona in 1921 and as seen in other places in Mr. Holland's papers it was sometimes a place without much tolerance and kindness.

The drilling began on November 1, 1920, and in three days the bit was already down to 97 feet. That would seem to be a good indication that Mr. Wammock was able to do his job. Soon after the hole is going crooked in the fissured limestone and he successfully gets it straightened by November 14, and continued down to 195 feet in just four more days of drilling. They were making splendid progress and it would seem that Mr. Wammock is doing an entirely satisfactory job. In just seven actual days of drilling down, he has penetrated 195 feet of rock with the churn drill. It is at this point that the gasoline engine fails and the drilling stops. This mechanical failure was not Wammock's or Holland's fault. If there was any fault, it was in the fact that stationary gasoline engines were only just past the experimental stage for heavy industrial work, and parts break. The breakdown prevented the drillers extracting the casing from the hole. By the time replacement parts arrived and were installed many days had passed.

The pulverized limestone and water in the bottom of the hole had essentially hardened like cement. After fighting for days to pull the casing out Mr. Wammock and Mr. Holland have no choice but to blast off the massive piece of steel on the bottom of the casing called the casing shoe. When they still could not extract the casing, they had to also shoot off the bottom 14 feet of the casing itself with explosives. Small and large pieces of metal created by this detonation would continue to be ground up by the bit for weeks.

Mr. Wammock's name is never actually mentioned in later criticism of the drilling disaster Crater Mining Company Hole No. 1 became. Holland and the next superintendent Mr. Plumb will later get much of the blame from the Barringers. In fact, none of these men were bad employees. The lengths and means that Mr. Wammock employed to fish out lost tools and stems speak to his experience and ingenuity. They often did not have the proper fishing tools on hand and so instead fabricated tools on site. Only on one occasion is it mentioned that they had no other choice but to order a special fishing tool and wait for its arrival to clear an obstruction. For the most part, he and his crew were able to snag and raise the heavy items lost at the bottom of the hole. Had the underreamer not broken into two pieces and been laying flat across the hole held down by the casing they might have been able to extract it as well. It had broken off months before, and they believed it was sidetracked outside the casing far above the bottom of the hole.

Several of the drill stems broke in a one week period. They broke at the same location on the stem regardless of the actual length of the stem. A huge problem, costing considerable time, which forced Wammock to use a drill string far too light for the work. The "standard" drill string employed a stem that was 30 feet long. Once all of those had broken, (and they always broke at the box on the end) he had to use the shorter ones of 18 feet. When those also broke he had to use stems of only 10 feet. Instead of crushing the rock with around 2500 pounds of tool string he was forced to use far less weight and make little progress. But he and his crew were always inventive and rose to the challenge to try and continue the work.

Holland had some pretty high standards that he was holding Wammock to. And this was to be expected for he was foreman and being paid the highest hourly wage on the site. He never mentions the persons responsible for the actual things that went wrong except on a single occasion where states that a bit was lost on Wammock's shift because it unscrewed. That was a serious preventable mistake. Wammock or his crewmen had simply not used enough force to tighten on the bit.

Holland is by March 11, 1921, dealing with a small rebellion at the camp. One of the men working on the drill crew has sent a tattle tell message back to California. Mr. Coan has become aware of dissension within the crew. In a letter of response to Coan, Holland brings up that he thinks Mr. Wammock is a good driller but not as qualified to be supervisor over other men. Holland sounds a little embarrassed that such tattling is going on behind his back. Here is what he writes in his letter back to Mr. Coan. "I have your letter of 5th March. I think your informant has exaggerated the delinquencies of the party you have in mind, though the fact that the information comes from one of his men is an indication that there was a lack of cooperation amongst the crew and scant respect for his ability. I think that, as often happens he fell down in attempting to handle other men though I have no doubt that he is a first class workman and would give good satisfaction if he had nothing more than the responsibilities of a driller for his own tour." Though Holland does not use Wammock name in his letter to Coan, he follows the quote above with a discussion of how the man brought his son as a tooldresser and the boy has difficulties. Clearly, this is Mr. Wammock. Holland says it is another reason for the men to have issues with the foreman. It is hard to imagine that the men would feel so badly about the man who had recommended them for the job. But, after months of struggle, they may all have gotten a little frustrated with the work. Perhaps they have begun to feel it would be better to be in California drilling at the Oak Ridge Oil field then fighting constant problems in the isolated camp of Meteor Crater.

The depth of the hole at the time it was declared lost was 312 feet. Just 117 feet deeper than Mr. Wammock had reached when the gasoline engine failed months before. He had reached the 195-foot depth in less than three weeks. With nine decades of hindsight, it is possible to imagine that if the engine had not failed they might have continued to drill down. They might have only faced the problem of fractured and fissured rock. Mr. Wammock might have been able to continue to straighten the hole when it became crooked. They might have even been able to reach the depth Barringer, and the bosses in Boston had planned. But Mr. Plumb's continuing problems for more than a year after Holland left make such visions of possible success for Holland and Wammock just what they are, dreams. Meteor Crater had nothing but pain and struggle in store for all these men.

Mr. Wammock's name is unknown regarding Meteor Crater. Until this writing, he has been forgotten along with Mr. Holland as even being a participant in the drilling program on the south rim of Meteor Crater. Mr. Holland's papers logically do not include anything after his resignation effective May 1, 1921. No record shows whether Mr. Wammock was the drilling foreman to the end of the drilling. There is evidence from D. M. Barringer Jr. in the 1964 biography of his father that the drill was changed to a rotary drill when work restarted after digging the tunnel. Did they change drill foreman and crew also? As with Mr. Holland, it seems too harsh to say that Mr. Wammock was responsible for the disaster this hole became or the tremendous expense it cost.

The Visitors

While there can be no story telling about Meteor Crater in the 1920s without including Daniel Moreau Barringer in the drilling operation on the south rim, he is only a supporting actor. It is his crater, and he is to profit richly if the drillers are successful in finding the buried asteroid. He is unquestionably very interested in the drilling work, but he is not in charge of that work, and much of his communication with L. F. S. Holland is about other issues. He has for years been inviting almost anyone that was interested to visit the crater. From mid-May of 1920 until the end of April 1921 Holland was their host, but Holland was responsible to men in Boston who had security and publicity issues of their own.

D. M. Barringer himself along with his son Reau visited the crater the last week of June 1920. On Wednesday, June 30, at 8:30 AM Barringer left Holland a neatly tri-folded letter with "Mr. L.F. S. Holland Personal" written on the outer flap. The letter was left where Holland would find it. The note was both instructions for what he wanted Holland to do for them that day, as well as an agenda of their activities. Reading hundred-year-old handwriting can sometimes be challenging, and D.M. Barringer's penmanship is often very difficult to read. But this particular letter may have been written at a time when he was relaxed and unrushed for it is quite readable. The contents of the letter are summarized as follows.

Barringer and son Reau have gone around the rim to the south slope to investigate the three newly dug shafts that Holland has sunk into the ejecta. Barringer tells Holland that they will not be returning for dinner and wants him to bring out their dinner. Barringer has asked the cook to make a meal for four. Holland and a guest are invited to join them for dinner wherever they end up in the evening. He hopes that Holland will be able to go with them to the dam to see the shotgun experiments. They will await Holland bringing the car to the south slope. Barringer asks Holland to bring the shotgun and pistol and all the cartridges that he will find just inside the door of the stone museum. He reminds Holland "to bring plenty of water for it promises to be a hot day and bring any mail and telegrams and a newspaper" to him.

If I worked for someone else and had my day already planned only to find a note like that on my desk first thing in the morning I might have some questions about what to do. Holland's entire day is now devoted to Barringer and his needs and requests. Mr. Holland has only been on the job at the crater for a month and may feel obligated to do all that Barringer wants. He certainly knows that his employers have a contract with Barringer. And he has likely been told to accommodate Barringer. But as politely as the letter is worded it seems a little presumptive on Barringer's part to so completely occupy another company's employee.

Arrangements for visits to the crater were usually made by Western Union telegram. And numerous telegrams with arrival updates and train numbers were received by

Holland. He was host to Professor Magie and his wife in July. This was an important visit as Magie was one of Barringer's most faithful supporters in the scientific community. They would arrive just a few days after Barringer and his son departed. It was hoped that his visit would help him to interpret the magnetic survey data. The plan was for him to see the measurements being made and have a better grasp of what the variations in the local magnetic field might reveal. Holland was a gracious host once again, and they had a very pleasant visit. Holland was given no instructions about restricting the information Professor Magie was told. But that was not always the case.

In late October 1920, Dr. Cambell and his wife visited the crater with Mr. Jennings' permission. Barringer instructed Holland that Jennings had agreed to their visit as long as they were not told much and left to wander around on their own. They were to understand that no publication of anything about the crater was to be made because of the work's commercial nature. Holland later writes that he did as instructed and let Dr. Cambell and his wife enjoy the crater on their own without being told much. Before the Cambell's arrival, there was a problem with their hotel arrangments. Ray Gebhart, the driver for the crater, was informed in Winslow at the Western Union Office of a message that was urgent, so he opened the telegram and read the Cambell's arrival date. He was not able to get them a room at the Harvey House, the customary hotel for crater visitors. Gebhart was able to arrange accommodations for them at the Winslow Hotel which appears to have been satisfactory. Holland writes Dr. Campbell to let him know of all this on October 28, 1920, and says that he will send Gebhart to "bring you and Mrs. Campbell out at your leisure. I can not come myself as we are having many little troubles with the new plant in starting it up." His wording is very nice but also served the purpose of distancing himself from direct contact during the visit of the Dr. and his wife. The statement was also true; they were having troubles. Holland wrote to D. M. Barringer about the Campbell visit in a letter the same day. His words are similar. "I will send in for Dr. Campbell in the morning early and hope he will depart early as we are having Hell with the gasoline engine and its accessories, and forgot how to be polite some days ago."

There is a backstory to the visit of Dr. William W. and Mrs. Campbell which began earlier in the year. Dr. Campbell was the Director of the Lick Observatory of the University of California. He had wished to visit in the crater in late May with George Ellery Hale the astronomer for whom the 200-inch telescope at Palomar is named, Paleontologist John C. Merriam, and John Casper Branner at the time President Emeritus of Stanford University. Barringer insists that their visit is postponed. He suggests late summer or autumn. Barringer was at a crucial point in his revived negotiations with United States Smelting Refining and Mining Company. Branner had just published an article about the craters on the Moon in which he favored the volcanic origin theory. Barringer was afraid that any publication that was not favorable to his ideas might again cast doubts in the minds of the men in Boston. Campbell and his friends must delay their visit. Campbell was very disappointed and complained strenuously with Barringer that the visit needed to be in late May. Barringer was having none of it. Campbell and his wife come alone in October. The ironic twist is that when he does visit and is told to write nothing without permission Campbell has no intention of writing anything. Campbell is also a supporter of Barringer's ideas without Barringer realizing it. Campbell seems to have little interest in aiding Barringer later after this lack of cooperation. An earlier visit by such prominent astronomers as Campbell and Hale could have benefited Barringer in future arguments within the scientific community.

During the winter of 1920, several of the officers of US Smelting Refining and Mining visited the crater. Holland was up to his neck in problems when Vice President Jennings was there on December 3rd. It was during his visit that they were planning to shoot off the bottom fourteen feet of the casing with explosives. It had to be a difficult time for the superintendent. The vice president of the company, his direct boss, is there watching him during such disastrous times. Jennings' later correspondence does not seem to reflect anything of a critical nature. Such events may have been just part of the drilling process and routine to men who have spent their lives doing this work. It appears from the numerous mentions made about drilling into creviced rock that it was known to be fraught with difficulties. Barringer discussed the problem long before the actual drilling began. But still, no one likes to have their boss watching while they are struggling at their job.

Jennings left the crater on Saturday, December 4. He had been there just a single day. Later on the 4th, they did shoot off the bottom fourteen feet of the casing with 23 sticks of dynamite. They had blasted off the casing shoe several days earlier. When the drill log is examined, the entry for all the blasting was listed as occurring on the same day. Holland's reports to Boston would indicate that the casing shoe and casing were destroyed with explosives on different days. Holland spent much of December 4 writing reports. He sent his weekly report to Mr. Anderson in Boston. He sent a report to Mr. Jennings even though he had been there the day before. Holland compiled a list of unused materials on hand at the crater which Jennings had requested the previous day. The long inventory sheets total up to $19,418.22 in equipment and materials and their associated freight costs. He had a busy day of paperwork on December 4th and was likely happy the boss was gone.

Mr. Patriquin of the Comptroller Department of United States Smelting Refining and Mining had been at the crater for a visit two days before Jennings on Wednesday, December 1st. Mr. Anderson the Consulting Mining Engineer of USSR&M Co. came out to the crater later. All of Holland's weekly reports were sent to Anderson along with Jennings. Holland writes on February 17, 1921, that he is meeting the train to pick up Mr. Anderson. At the time of his visit, they are deciding about the hole blocked by the underreamer bit. Again a bad time for a visit from another boss. Mr. Wammock, the foreman, declared there was nothing they could do, that the hole was lost. It would be decided quickly by Anderson and Jennings to dig the tunnel to clear the hole. Twice during the week of November 27, 1920, the crater was the host to the county supervisors, and the county and state engineers. They were there to decide on the route for the new road to Winslow. The three existing roads were atrocious and always in need of repair. Holland lent them roustabout Terrell who was thoroughly familiar with the area to show the men around as they planned the route. By the end of the year, the new road was scraped for fifteen miles, and improvements had also been made to Chavez Pass Road and Coon Tank Road which were to be part of the new road to Winslow. After the roadwork, Holland says it was still heavy going but much better than the old roads. The Overland car and Ford truck were often broken down because of the roads. At the end of 1920, the Overland car was out of service most of the time. The old highway to Winslow was described in a 1920 article in the Chicago Tribune as "the worst stretch of road in the whole United States."

Mrs. Holland was at the crater for a visit during the winter of 1920. Mr. Holland wishes that his wife could have remained long enough "to see the fine weather of January" that came after she left. He speaks of frost at night but beautiful clear daytimes. Holland is hopeful in a letter that he can have her return again soon. But within just days the hole will be lost, and the tunnel digging will begin and his time at the crater will be short.

In what may have been his last correspondence with D. M. Barringer, Mr. Holland writes on April 23rd that he has had a Princeton graduate named Sinclair Armstrong at the crater who came accompanied by a letter from Dr. Magie. Mr. Holland was quite impressed by the man. Sinclair spent all the time that Holland thought was good for him working in the dusty tunnel. The young man seemed to like it. After that, he began riding for Mr. Hart's outfit until leaving for Wyoming.

As isolated as Meteor Crater was in 1920 it was visited by guests and the curious on a regular basis. Since Holland was in charge, he was host and head of security. In the early years after Barringer acquired the crater, he had encouraged any scientist interested to come for a visit. He had built the original stone museum on the northwest slope and placed in it specimens of all the rocks that he had discovered proving its impact origin. Once the site became a commercial operation in 1920 access was restricted and approval for visits more commonly needed. By that time the debate over the crater's origin had escalated and drawn in many scientists on both sides. Added to the debate was the issue of the size and survival of the asteroid. This portion of the argument directly related to the work of United States Smelting Refining and Mining Co. and Barringer. They wanted as little discussion about the asteroid and its size and weight as possible. To limit that discussion visitors were screened and publication rights were restricted. Barringer's son was told to hold up publication of a scientific paper he had written, and Holland decided not to accept an offer to write a magazine article. The prevailing attitude amongst all the principles of this operation was "let the workers, ranchers, and visitors think we are foolish and let them laugh." The plan being the less they knew of the truth, the better.

The Drilling Crew

In all of the papers of L. F. S. Holland during his year at Meteor Crater, he never names any of the men in the workforce other than the drilling foreman Mr. Wammock and the roustabout Terrell. Holland makes mention of one of the tooldressers being Wammock's son, but we never learn his first name either. Ray Gebhard was not really an employee and seems to have been paid on a trip by trip basis as just an errand boy. Later, a few years after the conclusion of the work at Meteor Crater Holland prepares his "Drilling for Meteorites" manuscript and in that document, he names some of the men he calls the Ventura Boys. We know that they were provided by Mr. Coan of the Oak Ridge Oil Company. It was unclear at the beginning of the research how much could be learned a hundred years later about these men. For some Holland only gives their last name.

By 1920 there was a fast-growing oil industry in California. Many parts of Southern Calfornia were covered with derricks. Oak Ridge Oil Company discovered oil in 1916 in an area of Ventura County north of Los Angeles. The oil company took its name from the geology of the location the Oak Ridge Fault area of the mountains near Piru, Santa Paula, and Fillmore. The wider region around the original Oak Ridge discovery would soon be under development as a larger oil production area. Oak Ridge Oil's first well sunk in 1916 produced just 25 barrels per day and took oil from a sand layer at 3000 feet deep. Soon oil was being found at much shallower depths. Santa Paula Oil Company brought in a well at 2000 feet which produced 100 barrels daily shortly after the initial well of Oak Ridge Oil. The area still has some oil though the levels of petroleum recovered are now much lower then one hundred years ago.

As of December 31, 1916, Oak Ridge Oil had land holdings in Ventura and Santa Barbara Counties totaling nearly 5,000 acres. Most of the land was on a fee title basis, but a smaller portion was operated as a 1/3 royalty lease contract. They would drill more than 200 wells in the Oak Ridge Oil Field. By the time the drillers were sent to Meteor Crater, Oak Ridge Oil had become part of Ventura Consolidated Oil Fields along with Montebello Oil Company and other oil producers. It is interesting that in Holland's manuscript Drilling for Meteorites he mentions the close connection of Ventura Oil to United States Smelting Refining and Mining Company. In a newspaper stock market article of the time, the address of the General Offices of Oak Ridge Oil was 55 Congress Street, Boston, Mass. the same address as USSR&M Co. Ventura Consolidated was also incorporated in Maine as was USSR&M Co. It is unclear if Ventura Consolidated was an actual subsidiary of United Smelting as a host of other companies were including Crater Mining Co. We do learn from Holland's papers that he had a friendly acquaintance with Mr. Coan. But, others in Oak Ridge Oil in Boston may have been influential in the decision to use Oak Ridge Oil as the source of the men for the drill crew. Mr. Coan may have been instructed by his bosses in Boston to select a foreman and crew.

Iverson "Ike" Wammock, the foreman at Meteor Crater for the drilling, was born in

Pennsylvania in 1871. Based on some reports he was brought to California by either Standard Oil or Union Oil as an oil rig worker. He is listed as Crew Chief in one report. By doing just a little math it can be determined that he came to California between 1904 and 1914. His youngest child Virgina was six years old in 1920 and was born in California. Both his sons were born in Ohio and the younger of the two was fourteen in 1920. Holland mentions in Drilling for Meteorites that Wammock had recently survived a fire at Oak Ridge Willard #11. Oil well fires remain dangerous events but in the 1920s would have been very hazardous. In the early years, the oil pressure was high in the Oak Ridge field, and so the danger of blowouts was also high. As pumping and drilling continued the pressure in the field decreased and soon the newly invented blowout preventer systems would begin being installed. Mr. Wammock continued to work as a foreman and driller until at least 1940 where the census shows that he worked 52 weeks in 1939. Iverson Wammock died September 1, 1943. His wife Blanche died about 1960. The Wammocks resided in Fillmore, California at 445 Mountain View in a Craftsman Style Bungalow. The home is called the Iverson Wammock Residence in Fillmore Historical Society documents. The house was just recently built when he came to live there. The census records indicate that they owned the home. In 1930 he claimed a value of $6500 for the home and reduced the amount to $5000 by the 1940 census. Mr. Wammock was not highly educated having only completed the fourth grade, Blanche only completed the sixth grade, but we know from Mr. Holland's reports that he was an ingenious individual and as far as drilling was concerned very skilled. Holland also reports that Wammock was a physically strong man.

Howard Wammock, foreman Wammock's son, came to Meteor Crater as a tooldresser and Superintendent Holland reported that he had some problems there with his habits and manners. Howard would appear in several of the documents written by Holland because of these problems. Howard Wammock was only 15 years old when he went to Meteor Crater. He must have been a strong young man for tool dressing was not easy work. Tooldressers were paid well, and at the age of 21, he married. By the 1930 census, he was already divorced and living again with his parents. His occupation and place of employment in 1930 are shown as "driller oil field." As far as his personal issues and bad habits that the other men complained about when he was at Meteor Crater some of that might be because he was the youngest of the workers. But by fifteen he should have been mature enough to take care of himself appropriately. Leading to the idea that he had some difficulties of other kinds. Holland used the phrase "low intelligence" to describe young Howard. It is possible he was developmentally disabled, but that cannot be known for certain.

Holland uses the word "amongst" when he writes the list of the men in his manuscript Drilling for Meteorites. There may have been additional men at Meteor Crater. Some men may have come and then left, and it is possible the individuals he lists were those that stayed there until he left himself at the beginning of May 1921. Or this handful were ones he knew and favored most. Whatever the truth is, this can not be considered a complete list from his use of "amongst" in his statement.

Only the names of the members of the Ventura Boys are preserved by Holland. No further details are given other than that one man was killed in an accident near the time he wrote his manuscript. And he writes that the same man was pictured holding a mountain lion caught at the crater in a photograph. For some of the men only their last names are given, and it was expected that poor results would be gained from research to learn more about these men. However, it proved to be quite a different outcome. The online resources available now yielded considerable details on most of the men.

Les Currier, Bill Snow, the DePriest Brothers, Buckman, and Walter Linville, were the members of the crew which Holland innumerates. Research is a fascinating activity. One day you will sit down and hours will pass, and nothing will be learned about the topic you began searching. However, you will be taken on a ride around the past that leaves you knowing far more about something. On another day the files will fall together as if by magic and everywhere you look will be material on the subject you are investigating. Two days were spent researching the names which Holland had written in Drilling for Meteorites with nothing being learned. A month of other writing and work returned me to the research and that day was amazing. However, one member of the crew is still blank with nothing having been discovered about him, that is Bill Snow. There were individuals with the last name Snow living in the Fillmore and Santa Paula area where most of the other men were located, but none of the correct age had any connection with oil work. There were also Snows in the Los Angeles area but again no indication that they were involved in oil well drilling. Bill Snow will have to be the research project of someone else in the future.

The Buckman of Mr. Holland's list was also a very difficult individual to research after nearly a hundred years. There are strong indications that the following individual is the correct person, but it is not possible to eliminate all doubt. Ross Buckman was listed in the Historical Records of Fillmore, California as having resided at 417 Clay St. and that he was a watchman and tooldresser. He was born in either 1881 or 1882 in Maryland. Both dates are reported. He was 38 years of age in 1920 when the work at Meteor Crater was done. This supports the 1882 date of birth. He married his wife May when he was 21, and they had two children Ethel and Marie. They rented the home in Fillmore for $25. By 1930 his occupation shown on the census form was "tooldresser oil field." There was no entry for him or his family in the 1940 census of Ventura County. There were no other Buckmans or Buchmans in the area at the time which appeared related to oil field work. Fillmore and Santa Paula were the largest of the communities near the Oak Ridge Oil Field, and they were communities where large numbers of the petroleum workers lived.

"The Depriest Brothers" is all Mr. Holland writes, so it is not even known how many brothers there were. It is easiest to think of two brothers, but it is possible that there might have been three or more. Oil well work was one of the major employment opportunities of the time in Ventura County. It was more common for the men of a family to all work at the same job and with the same employer nine decades ago then it is today. A search again of the census records for Santa Paula, California where the offices of Oak Ridge Oil were located brought up a 1920 record for Walter D. Depriest. He was born in New Mexico in 1892 and was 28 years old. His occupation and place of work were listed as Rig Builder Oil Company. His father is recorded as being born in Louisiana and his mother as born in Arkansas. His living in Santa Paula was a good indication that he might be one of the DePriest Brothers, but more information was needed. Where was the other brother or brothers? An expanded search of DePriest names in Southern California revealed L. V. DePriest living at 221 So. Friends Ave., Whittier, California. L. V. was born in 1890 also in New Mexico; his father was born in Mississippi and his mother in again Arkansas. Mississippi and Louisiana could be a mistake in knowledge by the brothers. His occupation and place of work were listed identical to Walter's as Rig Builder Oil Company. It was looking good that they were brothers. But it was necessary if possible to find just a little more to settle the birthplace of the father discrepancy. A check of 1900 census records revealed a Warren E. Depriest born 1882 in New Mexico. His father was born in Mississippi and his mother in Arkansas. Further searching found the parents of Warren were James. P. Depriest and Amanda I. Depriest who were discovered to be living in Santa Paula, California on Ventura Street. A listing of their children still at home on a later census after the three boys left home, showed a daughter Mabel. The connection of the Whittier California Depriest family and the Santa Paula Depriest family was nearly fully established. But it further helped that the 1920 census record for L. V. showed a sister named Mabel living in the home in Whittier. The two "Rig Builder Oil Company" workers were brothers and with little doubt were sent to Meteor Crater in 1920 to drill for a lost star. What about Warren the possible third brother. There is no record that he was a well driller or tooldresser. But L. V. DePriest's first born son was named Warren, so little doubt remains; he was a third brother.

Les Currier was a surprise. Searches for Les revealed nothing, and neither did searches for Lester, then out of the blue, a census record for George Currier showed a Leslie Currier as a son. The 1920 census record showed he was 27 years old having been born in 1893 in Maine. He is the only one of the crew that lists his occupation as Driller Oil Wells. He was one of seven children, five girls, and two boys. His younger brother Frank was 26 years old and listed as a Tooldresser Oil Wells. Frank is living in the bunkhouse community at the oil field. The census taker has had to devote two pages of the book to the employees of Oak Ridge Oil Company that live up in the mountains in bunkhouses and homes. Leslie and Frank must have been close brothers. It seems likely they spent time together when not working for just a short time after the drilling at Meteor Crater they married a pair of sisters. Leslie Currier married Vina Jane Thomas, and Frank married Zola Belle Thomas. It is not hard to visualize the brothers double-dating the sisters. Frank married first on January 26, 1922, and Leslie married on July 6, 1922. The drilling work was in trouble during the summer of 1922 at Meteor Crater, but the drill crew was not dismissed until November 15, 1922. This might be an indication that some or all of the original crew of drillers named by Mr. Holland were not brought back after the digging of the tunnel when they changed to a rotary drill.

Leslie was in Santa Paula getting married during a terribly difficult time at the crater. They were down to 1300 feet, close to the final failure that would stop the drilling forever in August 1922.

One of the members of the crew gets a few extra words written about him by Mr. Holland in Drilling for Meteorites. Walter Linville is said to be pictured holding a wildcat that had been caught at the crater and there followed a discussion about the massive numbers of rattlesnakes that were killed during the work. Holland also writes that Walter Linville had recently (at the time he presented Drilling for Meteorites) been killed in an accident. This was not the first mention of Walter Linville and an accident. Holland does not say he died in a car accident; it could have been a workplace or other type accident. But the first reference to Walter Linville is regarding a car accident in which he and other workers at Oak Ridge Oil were involved. The accident made the local newspapers, and the following is the short article of May 16, 1919, from page six of the Oxnard Courier.

"An excursion to Hueneme by four Santa Paula men last night, for the purpose of indulging in "moonlight fishing," resulted in serious injuries to several of the merry group, on account of an auto accident. It was a Dodge car, driven by Harry Doboo. The time was about 10:30 and the place near the Wiltfong corner on the Hueneme-Oxnard highway. With Doboo were Pat Smith, Walter Linville, and George Chavez. They are all employees of the Oak Ridge Oil Company of Santa Paula. They were driving on the highway Doboo says, when suddenly there appeared another car, from one of the cross roads, directly in front of them. Doboo says in order to avoid a collision, he swerved his car to one side, turning turtle. Doboo, the driver, was lucky and escaped being hurt. Walter Linville was the next luckiest. He suffered some abrasions on his face. George Chavez fractured his left collar bone, and Pat Smith suffered most severely. His left arm was badly mutilated as a result of which the third and fourth fingers of his left hand had to be amputated at St. John's hospital, where the party was taken after the accident. After receiving treatment the men were driven to Santa Paula to their homes, except Smith, who is still a patient at the hospital."

Walter Linville was one of the younger members of the crew at Meteor Crater; he was 18 years of age. Born in Indiana in 1902 he was single and listed on the 1920 census as a Tooldresser Oil Wells. He was one of the many men and families that lived at the bunkhouses in the mountain by the Oak Ridge Oil Field. There is some evidence to suggest that his older brother was also an employee of Oak Ridge Oil Company. There was a George R. Linville also living in the bunkhouses. Though he is shown with a wife named Marguerite J. Linville. Families were housed at the mountain community as well as single men. George's occupation and place of work are recorded as Foreman Oil Wells. George is 35 which is somewhat older than Walter, and it is possible he is his father but for Marguerite to be his mother would require that she gave birth at the age of 12 or 13 as she is but 30 in 1920 and Walter is 18. Not impossible but again a detail that may never be sorted out. It is more likely that their parents had other children and that Walter and George are siblings widely separated in age.

All the men on the crew could read and write, all were males, and all were white. As mentioned before this was a period in American history where economic conditions were very tough, and opportunities were not available to all members of society equally. Holland has expressed some prejudice about groups of people elsewhere in his writing, and the evidence of who went to the crater is in line with those views which were so prevalent in America. In fact, every single employee of Oak Ridge Oil Company and every spouse and child of an employee is listed in the census of 1920 as white except the two janitors who were Japanese immigrants who had not become naturalized citizens. Holland was in the same catagory, an alien British citizen who it seems never became an American citizen.

One other member of the crew is easy to forget. He is not on the list that Mr. Holland includes in Drilling for Meteorites. He is Mort Branson the first drill foreman. He had arrived at the crater before his crew was to come there, he was struck with paralysis in the night and evacuated to Los Angeles by train the next day. We know from Holland's reports to Boston that he follows up on Mr. Branson's condition which improves greatly after returning to his home. The paralyzed Branson was met at the train by his wife who the census records as Della Branson. Mort was 44 when he went to Meteor Crater. Described by Mr. Holland as "a dead one" he must have already been suffering from some illness when he arrived. Holland put him on the train for Los Angeles after getting advice from a doctor in Winslow. Los Angeles was a generic term for the west coast of Southern California. Mort Branson, his wife, and brother in law lived in the mountain bunkhouses of Oak Ridge Oil Company. Della may have met the train at Union Station in Los Angeles, or it could have been somewhere closer to Ventura County and their home. The brother in law was named William Lucas, and he was also an employee of Oak Ridge Oil, listed as a tooldresser.

The 1920 census shows Mort Branson as a Foreman Oil Wells which is exactly what it should. However, in all the entries of the two pages containing Oak Ridge Oil Company workers and family members only two of the men were listed as Drillers. The vast majority of men working on the rigs were listed as Tooldressers. There could not have actually been that great a need for tooldressers at the oil field. The company would have had a centralized area for sharpening and resizing the bits. Spare bits would have been made available at the rigs for changing as needed. What reason the men had for choosing to tell the census taker that you were specifially a tooldresser is unknown. Most of these men were likely doing the work of drillers, rig builders and whatever else needed to be done. Many other job titles are listed in the census including pipefitters, gaugemen, teamsters, cooks, bakers, carpenters, painters, cement finishers; the list goes on. But just two drillers and nine tooldressers are recorded. Additionally, from the small group of Holland's crew that lived elsewhere in the cities, there were several more tooldressers and "Ike" Wammock a Foreman.

The crew at Meteor Crater was a mixture of different ages from teenagers to middle-aged men. They were hard-working men used to heavy labor. They were also used to living in a bunkhouse situation. Most were doing so at Oak Ridge Oil. Some had wives and children, and some were single or engaged. They were certainly more isolated at Meteor Crater than in Santa Paula or Fillmore, California. The Oak Ridge Oil field was very different geologically from Meteor Crater. The Oak Ridge Oil Field was oil bearing sands squeezed into pockets by coastal faulting and folding of rocks. The rig placement was perhaps familiar to some of the men. Many of the early wells at Oak Ridge were on slopes and mountaintops. But, the equipment used in California was clearly stronger than that originally supplied to Holland. He would try to upgrade to the California tools but would never have time to use them. The rocks at Meteor Crater were far more fractured and harder in places than anything the men found at Oak Ridge. Neither the men that worked with Mr. Holland and whose history is recorded here nor the men later working for Mr. Plumb reached the depth goals of United Smelting Refining and Mining Co. or D. M. Barringer. The Ventura Boys all worked after their time at Meteor Crater. Oak Ridge Oil Company became part of Ventura Consolidated Oil along with Ventura Refining. It would be acquired by California Petroleum or as it was called, Calpet. It was to Calpet workers that Holland presented his manuscript Drilling for Meteorites. Calpet would be acquired by Texaco later, and still, later much of the region of the Oak Ridge Oil Company became a subsidiary of Occidental Oil. The wells are still there, maybe different ones today. The production is a trickle compared to one hundred years ago. Some wells only produce 4-5 barrels per day currently. The vast reserves of natural gas are gone used in the making of gasoline in the early years. Like for Mr. Holland, the months at Meteor Crater were just a short bump in these men's careers that last for decades. Without Holland's documents, none of these individuals would be remembered for their attempt to locate a buried asteroid.

The Company

How L. F. S. Holland was selected to become the superintendent of Crater Mining Co. is a mystery. We do know that he always made his contact information available through The Mining Journal. Whenever he was not working, he would place an entry in the magazine that he had returned to Hollywood. Crater Mining Company was formed as a subsidiary company of United States Smelting, Refining and Mining Company which had operations all over the United States and Mexico. They may have had some interaction with him in the past. It is clear that Mr. Holland was a new employee for he is let go as of May 1, 1921, in favor of employees with longer tenure. At least that is the reason he was told.

What is the story of United States Smelting Refining and Mining Company? First off they were jokingly referred to in mining publications as the company with the longest name. They were called within the industry simply U. S. Smelting. The name was often abbreviated with the letters USSR&M Co.

USSR&M Co. was organized in 1906 to acquire U. S. Mining Company which was not a long established company itself having only been organized in 1899. But this was the beginnings of what was to become a huge force in the metals industry of America. The company would have mining interests in Utah, California, Arizona, Nevada, New Jersey, Indiana, Kansas, Alaska, and Maine. They would also have mines outside the United States, especially in Mexico. Called one of the Boston "coppers" in the press they were by no means only interested in copper. They would be prominent in other resources such as gold, silver, lead, zinc, iron, coal and oil as well as their enormous copper activities.

On March 23, 1916, their stock began being sold on the New York Stock Exchange. In an article from the New York Times of that date, they are called the second largest smelter in the nation. Before this date, the company's stock was dealing on the Boston stock-market. The initial stock offering on the New York Exchange was $24,317,500 of 7% cumulative preferred and $17,555,750 of common stock.

During 1911 and 1912 the company began buying coal properties in the Carbon and Emery counties of Utah. Among the first acquired were the Black Hawk Coal Mine and the Castel Gate Coal and Coke Company which became the Panther Coal Company. They later bought Consolidated Fuel Company the largest producer of coal in the state. Fifty-two percent of Castel Valley Coal Company was purchased as well. The coal holdings in Utah of United States Smelting Refining and Mining Company were organized into The Utah Company on January 12, 1912. USSR&M Co. also had control of the Utah Railway that connected their mines to the Denver and Rio Grande Western Railroad.

Smelters require other materials in addition to the ore. The tremendous amounts of required fuel were taken care of by the company's coal mines, their need for fluxing supplies was provided by a lime quarry at Topliff, Utah which was located on the San Pedro, Los Angeles, & Salt Lake Railroad line. This mine supplied all of USSR&M Co. requirements for lime and provided a surplus which was sold to other smelters.

By 1917 United States Smelting Refining and Mining Company was a driving force in the metals industry. The following quote from "Copper Club and Mining Outlook" will offer a glimpse at the role they have already played in American mining. "Probably no mining company of the world has had more to do with the development of the metal industry than the United States Smelting Refining and Mining Company."

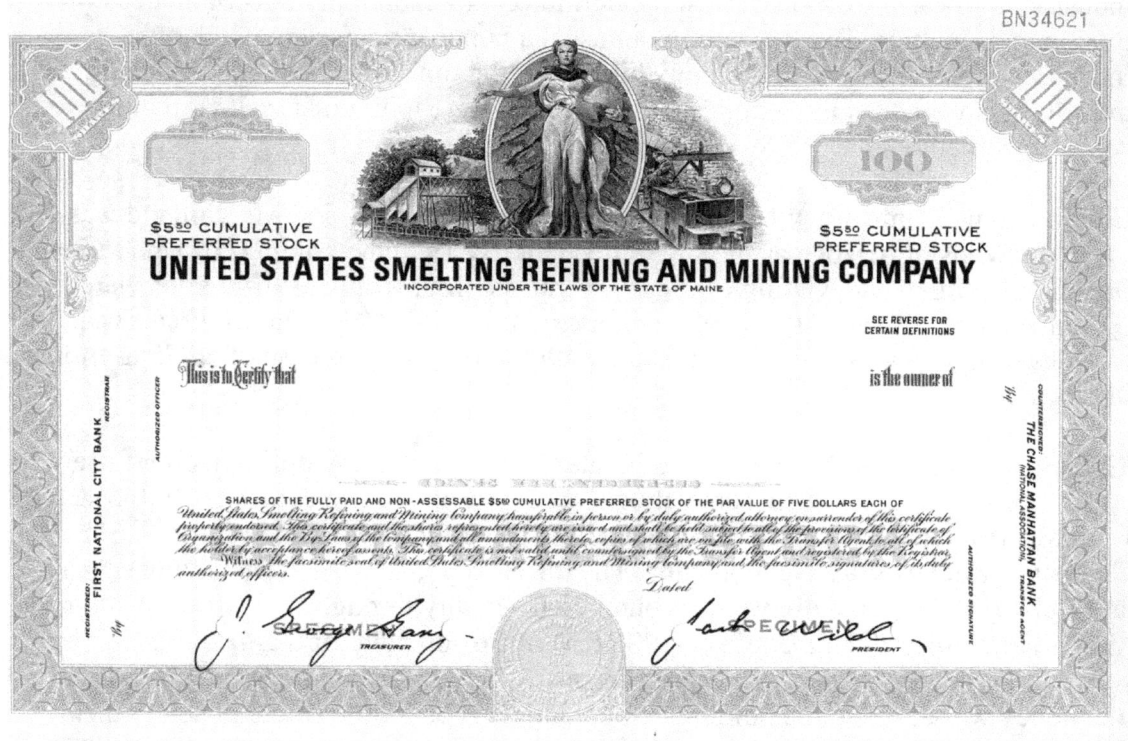

At this point, they are the holding company for subsidiaries all over the country. Among these subsidiaries are U. S. Smelting Company, Centennial Eureka Mining Company, Mammoth Copper Mining Company of Maine, Gold Road Mines Company, The Needles Mining and Smelting Company, Richmond-Eureka Mining Company, and many others. The host of coal mines, the lime quarry, U. S. Stores Company from which Mr. Holland orders much of the supplies for the operations of Crater Mining Company were also subsidiaries of USSR&M Co.

Mr. Holland writes many of his reports to Mr. Sidney J. Jennings the Vice President in Charge of Explorations. Mr. Jennings has enormous experience in mining and mine management. He has been a Vice President since returning from nearly 20 years in South Africa. Jennings was a consulting engineer for Boston Consolidated for two years before joining USSR&M Co.

Today it might be difficult to do what United States Smelting Refining and Mining did a century ago. The government would likely intervene and break them up as has been done with other huge companies. It is one thing to own the mines; it is another to also own the smelting and refining, the fuel and refining chemicals, and the railroads for the transportation of all the materials. Add to such an immense corporate structure company stores where the workers are almost forced to buy their supplies, and the word monopoly comes to mind. It would be impossible to operate such a corporation today. The antitrust laws would prevent them from controlling so many aspects of a single industry. USSR&M Co. was almost the definition of a monopoly in certain areas of mining and smelting. But this was not unusual a hundred years ago. All the big industries in America were operating this way at the time. Oil, steel, meat packing, many others were also monopolies. Often with a single individual at the top of the company structure.

Crater Mining Company was organized as another subsidiary, and Mr. Holland was sent off to the wilds of Arizona as the superintendent. By 1920 the holdings of United States Smelting Refining and Mining Company were vast. Many of the subsidiaries were capitalized in millions of 1920 dollars. For example, USSR&M had acquired The Needles Mining and Smelting Company in 1909 through a purchase of all $5 million of its stock.

There is no doubt that the operation of Crater Mining Company during its brief life was a financial loss. But there are some almost humorous twists in the money trail. Much of what Mr. Holland bought for the crater drilling program he purchased from U. S. Stores Company which was a sister company. The microscope, the tires from Goodyear Tire and Rubber and much more were all purchased actually through U. S. Stores Company. Mr. Holland ordered the materials, USSR&M in Boston paid for them, and the money was sent to the subsidiary U. S. Stores Company in Salt Lake City. So weirdly, not all the money was a loss to the holding company in Boston a little filtered back to them through the bottom line of another subsidiary.

United States Smelting Refining and Mining changed its name in the early 1970s. The change was to reflect the very diverse product line that had developed in the half century following the time of Mr. Holland. The new name was UV Industries. From this point, the company spins downward out of control for a while. There are numerous acquisitions, and transfers of assets. Lawsuits were filed against new owners by the EPA and other agencies for the tailing piles abandoned on the surface for decades. The tailings were contaminating whole communities. Sharon Steel Corporation acquired UV Industries in 1979 and inherited these lawsuits. Sharon Steel Corporation eventually filed for Chapter 11 bankruptcy in 1987.

In a peculiar twist, Mueller Brass Co. which joined with UV Industries 25 years earlier reemerged when Sharon Steel was broken up in 1990. Mueller Brass had been making brass plumbing parts and refrigeration equipment. They ended up with a coal mining division, a short-line railroad, an Alaskan gold mine, and several other mining interests in Canada and the Western US. It would seem that they received the remains of The Utah Company, the Utah Railway, and what was left of the mining interests of the original United States Smelting Refining and Mining Co. Mueller Brass Co. would divest themselves of all the mining and resource related operations by the end of the 1990's.

While the official reason for the name change in the 1970's to UV Industries was to reflect the change in products which then included electronics and refrigeration, it is easy to believe there was a motivation to get rid of the USSR on the front of the abbreviated name. Given the world politics of the time, it must have been a concern with some at the company that it was bad for business.

Crater Mining Company was a short-lived subsidiary of United States Smelting Refining and Mining. It was organized in the early part of 1920. Mr. Holland was employed by USSR&M Co. for just days short of one year. He was followed at the crater by Mr. Plumb who served there from May 1, 1921, until all the workers were dismissed on November 15, 1922. He may have continued with USSR&M Co. at another property. While mining is always a speculative and risky business the work at Meteor Crater was even more so. Mining engineers and exploration geologists 90 years ago did not have all the tools available today to know what was buried underground. Mining companies had to put money up and try a location that was promising with no guarantee of success. In the case of Meteor Crater, some credit must be given to the salesmanship of D. M. Barringer for persuading Sidney Jennings that a body of nickel-iron was somewhere under the crater. Instead of spending $75,000 on ten holes drilled to undisturbed rocks they drilled just one hole about a hundred feet short of that level spending close to $200,000. The one hole consumed 2½ years in struggles and failures. In the end, they learned little more than what they knew before starting to drill. USSR&M Co. gave up their lease after about another year, sometime in 1923. Barringer swiftly formed another company to dig a tremendous shaft and tunnel from the south slope down and over to the mass he believed the drill found.

Crater Mining Company passed into history without actually solving the mystery of whether there is iron under Meteor Crater. The truth of what they drilled through remains unanswered for some investigators to the present day.

L. F. S. Holland served USSR&M Co. to the best of his ability and was criticized along with Mr. Plumb and the crews. He was one employee in what was a massive company employing dozens of mining superintendents. He is never mentioned by name by the Barringers in anything written later. The kindest reference made by the Barringers is that Holland was "an inexperienced superintendent who was replaced by another inexperienced superintendent." Mr. Holland must have left the crater with mixed feeling on May 1, 1921. He had not received his final check from USSR&M Co. as promised. He would have to figure that out when he got home. Also, he was unemployed. For that period in America, he did well working for USSR&M Co. He made approximately $4,000 in the year of his employment at Meteor Crater. That was about three times the average yearly income. He called his work at the crater "worse than hell for him." So leaving might not have been such bad news. Sidney Jennings offered to give him a reference and assist him in finding other employment, again not a bad thing given Jenning's sterling reputation. But Mr. Holland left without finding the meteorite, with the drill hole blocked by the underreamer, and the tunnel to clear the drill hole only half dug. No one likes loose ends. He had spent all of the money USSR&M Co. had committed to the project, and the hole was just 312 feet down. The company could not have been happy about that. Balancing everything it is easy to see Holland smiling at the train station as he heads for home in Hollywood. As for USSR&M Co., they would have another year and a half of hell ahead of them at Meteor Crater.

The Work

Preparing the Drill Site

I can not believe how little we get done some weeks. It seems like everything is a problem here. I make weekly reports back to Boston and sometimes it is hard to find something to say that is new. It is just the same litany of misery every week. But I can not let them know more than the facts, and I must put a positive face on my reports. I know they want an estimate of when we can start drilling or "spudding in" as we say in the trade. But I just do not know when all this preparatory work will finish.

The site they chose is at the very edge of the crater rim. In fact at no place around the crater is the wall as cliff-like as here. I tried to figure out any way to mount the drill rig without clearing the area, but nothing was safe. You just can not have any of the rig out over the open air or try to erect it on uneven ground.

The ejecta is easy for the men to move with shovels and wheelbarrows. But the red sandstone is hard, and we have to drill and blast it with black powder. They muck away the material dumping most of it over into the crater. The workers are one of the biggest problems. They do not stay. Every payday most of them quit. Only a few have stayed for long; one is the cook the others are mostly older. The younger men move on and complain about everything while they are here. It is too hot or too cold, or they don't like climbing in and out of the crater. The dust is more than annoying to them and that part of the laborer's complaining I understand. Most of the men out here in Arizona hate to work.

I hired 12 muckers to work at the rig site from an employment agent. Four were union men that demanded 75 cents an hour, two left sick, and one smashed his toe. One of the men was insane and had to be turned over to the sheriff after he caused trouble with a rifle. There are no men to find many times when I look. Yesterday I found just one man to hire in Flagstaff. From the pained look he made carrying his belongings to the bunkhouse I can tell that he is a no account. The shortage of laborers is so severe the government curtailed their work of trail building across the canyons.

We have moved 5,000 tons of rock to clear the site for the drill rig and are not done. The area will be 100 feet long and 40 feet wide when finished. We had to cut down 20 feet in places to get the area level. The Fresnos we borrowed from Mr. Hart do not work well in the material of the crater. So we are using other scrapers. It's been blowing the last few days. The men have great discomfort working in the dust when it is windy. So I guess I will have to report again that the site is not quite ready enough to wire for the rig builders.

The timbers for the rig arrived in San Pedro, and have been sent on. So all the materials for the rig proper should be here soon. However, only a portion of the casing has arrived. The rest has been promised for several weeks. The tools finally arrived, and that is good. We will be able to at least begin the drilling. But we need that other size casing. It could come while the rig is being erected. I have been working to figure out the most economical means of getting everything from Sunshine Station to the crater. We have used two transport companies; Matlock and Donnell. Both have trucks much larger than our specially geared Ford. But the cost seems high and buying a bigger truck may prove to be better than hiring vehicles and men.

The drill rig on the south rim of Meteor Crater was erected very close to the edge. So close that when difficulties with the hole got bad enough to prompt discussions about whether to move the rig a little and redrill they could not do it. What they contemplated was to move the rig with jacks five feet north. The bosses in Boston preferred moving the rig even closer to the crater edge. Even just the five-foot move would have put the mud seal part of the rig off into the air.

The rig site was a cleared and leveled area 100 feet by 40 feet. They had blasted out the red Moenkopi sandstone down as much as 25 feet at the inside edge to make the site level. In all approximately 5,500 tons of rock was removed by the end of the work. Most of the rock was broken up, scrapped, or blasted, then dumped over the edge of the cliff into the crater. A team of muckers was on the south rim for the better part of three months preparing the rig site.

Churn drilling requires a large volume of water. The only water available was in Canyon Diablo to the west where dams had been built. A 15,000-foot pipeline and pump would move the water to the south slope camp and rig site.

Holland states the ejecta was easy to remove, which makes sense; it was pulverized rock thrown out by the impact. But the red sandstone was a different story. He describes it as tough and hard to work down. The red bare spot where the rig stood is still visible, and the rim trail goes right passed the location. If one turns aside for a look, the remains of the work are still scattered about even after nine decades. A small diameter pipe sticking out of a piece of casing cut off nearly at ground level marks the actual spot of the drill hole location. Below the rig site inside the crater, and on outcrops of the south cliff lie timbers gray with age resting where they fell. It appears the derrick was disassembled after the work and most of the timbers were just pushed over the edge. Used lumber must not have had much value in 1923. Which is a little surprising. Holland had invested great effort and time salvaging the wood from the old buildings on the crater floor and constructed the wooden slide to drag the lumber out of the crater. The timbers and siding were repurposed at least twice from those buildings to save money. The cost of lumber was very high locally when the spiling was bought for lining the tunnel in the spring of 1921. Frustration, disappointment and the desire to just wash their hands and walk away seem to be the reasons the company had for disposing of the lumber into the crater.

Just a short distance downhill from the drilling site are three tailing piles and their shafts. Though it is not known for sure if these are the three shafts that Holland had men dig. These do seem to be in the correct locations and match his descriptions. Their purpose was to examine the nature of the ejecta and the dip of the rocks below the surface. Holland's shafts were dug on the MN line, and that is where these three are located. Satellite images show how the shafts line up. The pole of the marker flag placed on Barringer's MN Line is still on the low knoll next to the drill site. On June 30, 1920, during his visit to the Crater, Barringer made a point of examining the three shafts. The color of the tailing piles can be used as a rough indicator of the layers that were reached in the digging of each. The shaft highest up the slope is surrounded by a tailing pile of nearly all red sandstone debris; the next shaft is likewise almost only red Moenkopi sandstone. The third shaft on the MN line is mostly the gray and white ejecta material with just a small center of Moencopi red in the tailings. No worker would ever throw rock further than necessary, so the material is sorted in a pile by depth. The deepest material on the top of the tailings.

The drilling site is literally sits on the edge of a cliff. It was very dangerous, and the workers needed to be careful dumping their wheelbarrows. Though no men fell, they could have fallen the great distance to the talus several hundred feet below. The drill rig was at least partially if not fully enclosed in a building. Holland reported on October 15th near the end of the riggers work that they were "housing in the outfit." After the building was around the rig, it made the site safer, but men still worked near the cliff.

The mud from the drilling was run over the edge and down the face of the south cliff. To do the drilling the bottom portion of the hole was filled with water. The bailer would be sent down the hole where it would fill with the muddy mix of water and pulverized rock. It would be raised and emptied. Then lowered and filled again. This was repeatedly done until little water and mud remained in the hole. This mud could not accumulate next to the rig. It was dumped so it could run over the cliff edge. At certain times during the drilling, the material in the bailer was collected and tested for metal especially nickel. During the early drilling into the limestone, the material was not tested often. But later the testing was nearly a daily routine. Piles of bags of this sampled material would have been stored or left lying about the area. Deciding on how the material collected for testing would be stored was a big decision. Bottles and cans which Holland saved from the mess house were used at first. Then later a paper type box with a screw on cap was used. The contents of the bailer were dumped in a trough and allowed to settle. The water was poured off, and the sludge was then dried by leaving it in the sun. However, the winds were sometimes an issue, and the collected material would blow away once it was dry. Holland had to prevent this from happening. In the winter it would not dry at all, and the collected samples were taken inside to dry. Any bits of shaleball meteorite were separated if found. Barringer was fascinated by this material and would always want samples sent to him.

These are the two still standing buildings on the south slope of Meteor Crater. The large building in the foreground still had bed frames and a dilapidated wood burning stove in it when visited in 2000. The buildings are shown from the other direction in a photograph taken by Mr. Holland later in this chapter. They were newly built in his picture, they are in very poor condition now after weathering through nearly a century of storms and high winds.

Somewhere at the drill site at Meteor Crater, Mr. Holland had set up an office with one bed. Presumably, this was his dwelling and work space. He had his little assay office area too. He had received some of the reagents for the nickel test from D. M. Barringer. His reports and Barringer's letter do not say which of the chemicals were given to him. Dimethylglyoxime is the well-known reagent used in testing for nickel and some other metals, but for the test to be accurate and trustworthy, it required proper procedures be followed utilizing several other chemicals. Nickel is not alone in iron meteorites. Nickel is a few percent of the alloy the rest is iron with traces of several other metallic elements. One of these others is Cobalt. As it happens, dimethylglyoxime will also give a visual response to the presence of both iron and cobalt. The color of the red precipitant is different to the trained eye but can be a problem for the inexperienced amateur chemist. Cobalt is never a problem in the testing of meteorites. It is usually under a tenth of the amount of the nickel and usually below the level to show a response. But the iron is the predominant metal in meteorites, and it must be eliminated from the test. This is where some of the extra chemicals become important. Holland would have gotten all the reagents needed and glassware of some kind to do the analysis. He reports the results of the tests often as does Superintendent Plumb his replacement.

Where Holland does this testing is not mentioned. Holland's original layout of the camp described an office building with a bed for the superintendent. He had a typewriter and a microscope and the mentioned lab chemicals and glassware. He had to keep correspondence, receipts, and office supplies as well. Some furniture was purchased from the local cattle rancher a Mr. Hart. This furniture may have been placed in Holland's office.

Three Jones Samplers were at the crater to split the collected material into smaller samples for the nickel testing work. The machines had riffles to catch a specified portion of the material run through the unit. At a normal mine site say that of a gold mine, the process was to mix the rock from a cut taken in the mine and crush the material. Then mix it again. Often a timer was used to guarantee that mixing was always perfect. Then further crushing of the rock and mixing again. When the particle size was small enough, it was sent through the Jones Sampler, and a fraction of the material was caught for analysis. If the mixing was done correctly the refinement of the small sample in a furnace would perfectly reflect the amount of metal in the ore of the area where the original large cut was made. This procedure was performed day by day to ascertain whether miners were working in good ore or low-grade unprofitable ore. Holland had two 10 x 18 and one 8 x 10 Jones Samplers at Meteor Crater. By using the samplers in a group, it was possible to extract a fraction with any degree of reduction percentage desired. If one machine was used, it was often referred to as "quartering" the sample. By running that fraction through another machine it could be reduced to say a sixteen of the original amount. The built-in scoops which caught the material as it passed through could also be changed to catch differing amounts. The 10 inch and 12 ½ inch drill bits cutting through thirty or more feet of rock a day could generate a great amount of material. Too much to go through by hand. Some visual examination for specific items such as shaleball pieces, followed by the machine sampling of the batch was likely Holland's plan. The larger Jones Samplers were $30 each and the smaller one cost $20. They were bought on February 7, 1921, freight prepaid to Sunshine Station. They were of little value to Holland being bought while the drilling was just starting into the sandstone. It was also just precisely the moment that the hole was declared lost by the drillers. Somewhere around the rig site, the Jones Samplers would have been set up ready to process the collections from the drilling. Mr. Plumb likely found them useful after he took over the work. There was little reason to have the Jones Samplers earlier since the meteoritic material was not expected to be found in the upper limestone.

Holland used magnets to extract the metal from the samples as well. Since the beginning of the drilling, equipment was lost through breakage. They would try to fish parts out but when that failed they would end up being pounded on to pulverize them. Other iron fragments from the casing that was blasted were in the hole. The bailer continued to bring to the surface chunks of iron that was not meteoritic for months. Holland had only reached the bottom of the limestone when the drilling was stopped and digging the tunnel started. At 312 feet it was expected that there would be little meteorite material in the bailer. They anticipated finding it much deeper down in the gray and white sandstone. The evidence from the 28 drill holes into the crater floor done in Barringer's early years showed the meteorite fragments were found between 400 and 680 feet. The drillers on the south rim had to add the height of the crater wall and expected to find meteoritic material after the hole was passed 1,000 feet in depth.

Usually, a wheeled cart was at a drill site to move the drill bits to and from the forge house. Sometimes the forge was inside the building with the rig. It appears that at Meteor Crater the forge was in its own building. Just as a safety issue it was much less of a fire hazard that way. The rig and derrick were constructed of wood. Steel derricks had been available for many years, but wooden ones were still common. Drill bits could weigh hundreds of pounds. They were many feet long when new. A 10-inch bit similar to the one used by Holland was listed in the Fairbanks catalog of 1914 in three lengths, 5 ½ feet, 6 feet, and 6 ½ feet. The six foot long bit weighed 925 pounds. They would gradually get shorter as they wore off and were dressed. Dressing the bits required heating them to red hot in the forge and then hammering on a new cutting edge. Bits often needed resizing also after forging. Tools in the shape of a ring were used to gauge the size of the bits as they were dressed. The type of fuel for the forge is not mentioned by Holland.

Numerous other pieces of equipment would be around the site. Many different fishing tools, drill stems, jars, cable hooks and such. Along with the actual wrenches and tools used to screw together the drill strings. Some of this could be stored inside the building surrounding the rig. Other items were far too large.

There were also spare engine parts, two coils of 2 ½ inch bull rope, 30 barrels of gasoline, 4 barrels of oil, 3004 feet of 6 ¼" casing and 3017 feet of 6-inch casing listed on Holland's unused materials inventory of December 4, 1920. There were 3000 feet of 10-inch and 3000 feet of 8 ½ inch casing that was being used on site as well. The Overland Car and the Ford truck owned by Crater Mining Company as well as the vehicles of any of the workers had to be parked somewhere on the south slope. When the site was fully operational, it was busy and well-stocked.

There was doubtless somewhere on the property a dump. A place for the waste of the work and the mess house. Incineration was a common practice back then and though Holland makes no actual mention of how the waste was disposed of it is certainly possible that much was burned. This could have been part of the duties of the cook. Remains found on the crater floor near where buildings once stood give indications where dumpsites might have been for the early Barringer work. All the dump sites including the south rim site have been cleaned up or covered up enough that little is noticeable except to the person walking on the very spot. But people make trash and equipment breaks or is destroyed and someplace is usually designated away from other activities to contain the material. Glass bottles and cans, old drums and barrels, broken crocker all survive for years. They provide wonderful data for the investigator looking at the site from an archaeological perspective.

Holland took a few photographs of the site when the area was being cleared and leveled. He took a few during the work and took a few near the end of his term. He was not a good photographer and probably did not have much more than a simple box camera for taking his pictures. The fact that he had the pictures to include with his weekly reports might suggest that there was a photography store in Winslow that could develop his negatives. One of his photos shows the drill site being scraped and leveled and several workers are shown in the picture. A small knoll next to the site is clearly seen as it can be seen in satellite imagery of the south rim today.

The roads are likewise visible in satellite images, and they show us very clearly even after nearly a hundred years where vehicles were driven. Buildings remaining on the south slope of Meteor Crater today seem to be some of the ones built for the drilling program Holland began. The large building still had bed frames with steel springs and straps when visited in the summer of 2001. The beds were likely from the later activities such as the silica mine. Out the back of that building was a collapsed portion called the root cellar by a guide in the 1990s. The smaller building still on the south slope had a remaining pile of bags covering most of the floor. The purpose of the bags could have been for holding the silica sand during the Meteor Silica Corporation lease years or could have been Holland's sampling bags from years before. The old buildings stand but just barely. There are tremendous numbers of nails everywhere. Many times more nails than needed for construction. Holland had the walls lined with building paper to make them warmer and keep out the dust. Tar paper was not sold locally in 1920 and was known not to last well. The building paper was available, much less costly than tar paper and was suitable he felt for at least the first winter. Some of the nails might have been for repeatedly securing building paper. But the thousands of nails loosely hanging from the roof today may indicate the efforts taken to keep the metal roof from being torn off by the violent winds. So strong had the winds been at the old north camp that the roofs of the buildings were strapped down with cables and secured to deadmen mounted in the earth. The winds Holland reported on the south slope were no less violent. The drill rig and derrick were secured with 12 cables. No ropes secure the old buildings today. It is remarkable that they have stood against the winds that have lashed them since they were abandoned. On April 23, 1921, the most violent winds of Holland's stay blew down the old shaft house on the crater floor.

There is no water tank at the drill site today nor any remains of where it was mounted that can be easily found. It had to be higher than the drill rig and the buildings. Otherwise, there would be no water pressure at all. It was mounted 50 feet higher in elevation than the mess house and 30 feet higher than the drill rig. One site nearby that seems to fit that description is a knoll of red sandstone just to the side of the leveled drill rig site. The tank held 13,000 gallons and was moved from the old north camp to the south slope. The 15,000-foot pipeline connected to the tank. The pump at the lower dam out at Canyon Diablo kept the tank filled.

During Holland's time at the crater there would have been a location where materials for the second purchased rig were piled awaiting its assembly. Though much was used to supplement the shortage of materials for the first rig, a pile was somewhere at the site. Such a great pile of wood shows in one of Holland's photographs. However, it was a picture taken while preparation work was still going on and can not, therefore, be the wood of the second rig alone. The great coils of 2 ½ inch manila rope would be somewhere. Much of this material should not have been out in the weather. Mr. Holland makes no mention of any storage facilities on the south slope. No mention of tarps or wooden coverings for any of the supplies is mentioned. Preserving the unused inventory would certainly have fallen into his job description as superintendent. Mr.

Jennings on his one day visit to the crater December 3, 1920, requested Holland make an inventory of the unused materials at the site. His inventory lists $19,000 of supplies and equipment.

In the early days of the work, it was very primitive. There were no buildings at all. The workers camped out. Most of the men slept on the ground in the open. A few found room on the floor of Mr. Hart's bunkhouse and granary. The cooking was done in a small tent. Not a single cot could be purchased in Winslow, and it was three weeks until any could be obtained from elsewhere. Holland wrote that more than one prospective worker arrived, saw the conditions and drove away. He remarked that in earlier years the casual laborers in the west would carry their bedroll and possessions walking, but that by 1920 they drove their Fords from job to job. Some of the laborers even until the end of the pipeline work had to camp outside the buildings. Here again his reports show the negative side of American history and his personal prejudice. He hired Hispanic workers for the final work on the pipeline and made them camp outside. They had to supply and prepare their own food and promise to not go into the bunkhouse or mess building, for Holland felt they would offend the other workers. He held similar views and made similar remarks about the Native American labor in the area, though he never seems to have hired any Native Americans. His remarks about many of the white workers are very uncomplimentary was well. He refers to some as "poor white trash." Holland never seems to have become an American citizen. Census records of as late as 1940 show he was a resident alien from England. He held and expressed views that were sometimes very disrespectful of other persons. Whether these views come from his upbringing in England or were acquired later, they are an aspect of his nature that this author does not admire.

The camp buildings and both the Ford truck and the Overland auto are visible in this photograph taken by Mr. Holland and included in one of his weekly reports to Boston.

Little evidence remains at the drill site today. But enough tidbits of the work are there to keep anyone who is obsessed with Meteor Crater thrilled. Lengths of pipe and half buried parts of equipment rest where they were left. Bricks likely from the forge are seen in one spot. They were made by the Los Angeles Pressed Brick Company. If they are fire bricks from the forge, they would have been made in LAPBCo's Alberhill plant.

The Los Angeles plant did not produce fire clay bricks. The plant at 952 Date Street and Alhambra Avenue did manufacture the bricks for most of the important structures in Los Angeles of the time. Many of the buildings on the campus of USC were constructed of bricks from the Los Angeles Pressed Brick Co.

Timbers are scattered around the slope and vastly more are down over the edge of the cliff. One large pile of desert weathered gray beams rests at the top of the talus just under the site of the drill rig. Two things are very noticeably missing from the site today. Evidence of the huge cellar blasted out from beneath the drill rig cannot be seen, and nothing remains of the concrete mount for the gasoline engine which had required ten cubic yards of aggregate rock be purchased and delivered at high cost.

Much discussion went into picking the location for the first drill hole of Crater Mining Company. It was to be as close to the edge as possible and go down through the so-called MN line marked on the first magnetic study. The location was near the top of the "arch" of the uplifted south side of the crater. Anyone looking at that portion of the crater wall can see that the rocks are severely broken. The men in 1920 knew the rock layers were damaged. They expected to have to drill through crevices. From Holland's reports of the difficulties preparing the site, it would have been easy for his bosses to have sighed with relief that the riggers were finally coming to set up the drilling equipment. But if they did feel relief that the site was finally ready it was in vain for more serious problems were quickly in store for them.

The Water and the Pipeline

Since almost the day I got here, I have been a cannibal. I have been scavenging and consuming the materials of previous projects at the crater. Much of it is nearly twenty years old, rusted and not working. Now we need the old pipe from the crater floor, and anywhere else it is available. We could not get any pipe from Los Angeles or San Francisco. We got 10,000 feet of two-inch pipe from Pittsburgh. We connected it for a while to 5,765 feet of the old spiral riveted pipe, but we could not keep up with the leaks. So I had to find more pipe somewhere, anywhere. There was some in the crater floor and on the north wall. Much of this is no good either, but part of it can be salvaged. It looks like we can get 5000 feet. But it is in the bottom of the crater.

The slide is taking some time to build. The men hate to work in the crater. It is hot, and the wall is steep and climbing out of the crater is difficult. Men usually quit after just one or two days of working inside the crater. Most of the work on the slide is being done by just two men; Terrell our roustabout an excellent worker and a freak from a lunatic asylum. It will be complete in a few days, and we can pull the rest of the timber and the pipe out. I had it all moved by horse to the foot of the slide. We had hoped to supply the horse with water from the bottom of some of the old shafts. But the water is very foul and stinks. Likely from dead animals that have fallen in the shafts. The horse won't drink the water. The men are hauling water for both themselves and the animal while construction continues on the slide.

I decided since we will have the slide that it would be good to gather everything else of value from the crater and pull it out. There is not much else left, but there are some cables and dead men. We will certainly need the dead men for the buildings of the south camp. I've been told the old north camp buildings were often close to blowing away in the high winds. The only thing holding them in place and keeping the roofs on were the cables over them tied to dead men. Some of the massive iron meteorites have been used as weights to keep things from blowing away. The winds on the south side are at least as high as those on the northwest slope.

Preparations for the drilling program on the south rim included the creation of a whole new pipeline. From the pump station near Canyon Diablo to the 13,000-gallon tank at the south campsite was a distance of about 15,000 feet. It took most of the four months before building the rig to dig trenches and place the pipe. Some of the pipes were still exposed at the time the drilling began November 1, 1920.

The trench for the pipeline was dug through a mixture of hand labor and using a plow and scraper pulled behind a Case tractor. For a portion of the length, they uncovered the old pipeline and laid the new pipe next to it. This section was mostly rock and they obtained a more powerful truck and used a "rooter" to deepen the trench. The labor problems were so bad and men so scarce who wanted to work that Holland would have to move men from the rig site preparation work to the pipeline work and then back again often depending on the weather. If it was stormy and raining, they would work on leveling the rig site if windy they would work on the pipeline. Holland often had too few men to shuffle and so reported week by week little progress.

In another of Holland's poor quality photographs the steam powered Case tractor can be seen with three men on it and two men standing on a scraper to give it some weight for spreading the dirt. This image has developer stains and a large spot on the rear wheel of the tractor suggesting that Holland may have done his own negative and print processing. It is hard to conceive of a professional photography shop even in 1920 selling to a customer a print with chemical blotches which could be replaced with another correct print. Since this was his file copy of the weekly report, he may have sent a proper photograph to Boston.

The 13,000-gallon water tank was moved from the old north camp and set up 50 feet higher than the mess house and 30 feet higher than where the rig would be. But if they could not get the pipeline done and the pump to work they would have to haul water by mules from the north camp. They were still getting some water through the old north camp pipeline and had left that line intact. It had been keeping the tank full before it was moved.

The dams across Canyon Diablo were very poorly constructed and after the rains of August 1920 were leaking badly. Holland recommended if the work at the crater continued very long that concrete dams replace the earth, adobe, tar paper and brush dams that were being used. They were able to stop the leaking with many sacks of "dobe" placed at the base of the lower dam. The reservoir was full to about three feet from the top when that rainstorm was over. Unfortunately, there would be just one small additional bit of rain in October. By the time they had been drilling for a couple of months the prospects for having water enough for a longer drilling program were looking bleak. In fact, as the tunnel was being dug into the side of the crater to clear the drill hole Holland thought they would not have enough water for anything but the use of the mess house and the men.

Normally it is not permitted to just dam a river or stream without telling the authorities. One of the documents included with Holland's papers was a "Notice of Reservoir Site Location." The notice defined the extent of the reservoir area and the nature of the construction. It states that the dam will be made of earth, timber, and stone. The notice was to be posted on a stone monument at one of the corner locations stated on the form. The defined area was a square one-half mile on each side. Arizona was a territory on January 17, 1910, when the document was signed by a witness and the locator Henry R. Holsinger. Henry was the son of Samuel J. Holsinger, Barringer's foreman, the man who first told him of iron found around a hole in North Central Arizona in 1903.

There are two interesting statements in the reservoir notice that add some details about the activities of Barringer at the crater. In building the dams across Canyon Diablo a quarter square mile of land was claimed. This land is the only land out on the plains around the crater that Barringer seems to have actually had control of. The document also makes it clear with the words ". . .Public Domain subject to and under the jurisdiction of the Territory of Arizona" that the land was not owned by anyone but the people of Arizona. Mr. Hart the cattleman who was always helpful to Barringer and Holland had his own dam and reservoir some miles away near Sunshine Station so it can be guessed that these two dams were solely for Barringer's use.

There is often some hesitation seen in people seeking to get something changed with government agencies. It is often less painful to work with an imperfect situation rather than to alert the government to all the details of an operation to make improvements. It is possible that Holland with Barringer's permission could have gotten an amendment to this reservoir declaration so that the dams could be upgraded to concrete structures as he suggested to his bosses. But to do so would have meant getting engineers out to the property and much involvement with state and county officials. This could have led to the revealing of information about their work at the crater that they wished to remain secret. The result was they always had two leaky dams made of mud, scrub brush, rocks and tar paper. Nothing of those dams remains today in the windy narrow gorge for an explorer to find.

To be honest and fair a little clarification is offered here. In other places in this book it is written that Barringer never actually had claims for land where he collected meteorites. That out on the plains around the crater where the meteorites came from was not his property. It is clear he did control one-quarter square mile of land surrounding his dams out at Canyon Diablo. It is close to the crater so there likely were and still are meteorites on that piece of property. However, his map included in the early scientific papers showing where meteorites were collected on the more than one hundred square miles of land surrounding the crater predates the tiny land claim of the dam site. Barringer was collecting meteorites where ever they were seen, and that land was except for the dam area, not his. The land was public domain according to the Notice of Reservoir Location. Maybe this allowed for any use in 1900-1920 including meteorite collection. Meteorite hunting was allowed in the area until into the 1960s with just the request that the holes dug be filled in for the safety of the cattle grazing the area. Meteor Crater being a National Natural Landmark in private hands as a money making business has drawn criticism from some quarters in the public arena. But such

NOTICE OF RESERVOIR SITE LOCATION
--))0((--

TO ALL WHOM IT MAY CONCERN;
Notice is hereby given that the undersigned has this day located under Title 63, Chapter 1, Revised Statutes of Arizona the _Midway_ RESERVOIR, situated about _15_ miles in a Southerly direction from Canyon Diablo Station on the A. T. & S. F. R. R. and on Canyon Diablo Canyon in the County of Coconino Territory of Arizona.

That the undersigned intends to commence the construction of a dam within a reasonable time from the date hereof for the purpose of impounding water for mechanical, irrigation, domestic and stock purposes: that said dam will be constructed of earth timber and stone: that in order to protect the undersigneds rights to said water and reservoir and the full enjoyment thereof _we_ claims all that land described as follows. Commencing at a stone monument where this notice is posted and running West _2640_ feet; thence South _2640_ feet; thence East _2640_ feet; thence North _2640_ feet to the place of beginning, the corners therof being marked with stone monuments.

That the undersigned propose to mantain said reservoir for the beneficial purposes aforesaid, keep the same in proper repair and comply fully with the laws of the Territory under which such reservoirs may be constructed upon the Public Domain subject to and under the jurisdiction of the Territory of Arizona.

Posted this _17th_ day of _Jan_ blank 19_10_.

Witness _H. W. McCulloh_   Locator _Henry R. Holsinger_

sites are of a voluntary nature and either the government or the private owner can remove or withdraw the site at any time from the NNL program. The closing of the property to hiking and exploration has also drawn criticism. The property has been well maintained by the current owners, and the Crater has been preserved. The perimeter trail is in spots near the edge of the rim, and there is no fence to speak of. The temperature can often be high, and not all hikers come properly prepared. The liability of letting the public run loose on the property is understood, and damage would be done to the crater by millions of feet wearing down the land.

There was a problem with animals using and polluting the water at the upper reservoir. Holland feared that releasing any water through the upper dam would then pollute the lower reservoir. The lower reservoir was the supply for the camp as well as the drill rig. The rancher in the area a Mr. Hart had moved his cattle by the time the water situation got severe. He was in the process of moving his horses when the tunnel was being dug in the early spring of 1921. The winter of 1920 was remarkably dry.

In Holland's weekly report of August 10, 1920, he first mentions the creation of the 1200 foot long slide. He secured a horse whim from a local quarry to haul the pipe and lumber up the slide from the interior of the crater. By September the wooden slide was completed, and they were pulling the lumber up. But a casting in the horse whim broke. It could not be repaired or remanufactured locally. Even the repair facilities of the railroad shop were not able to help them. The new part had to be obtained from Albuquerque, New Mexico. On September 4 he reports that the new part has arrived and the whim is working. They were again pulling up the lumber and would pull the pipe up soon. September 11 he reported that the pipe was nearly all out of the crater. It was being distributed next to the old line. The crater salvaged pipe would be connected to the 10,000 feet of new pipe already installed. In this report, he also mentions the removal of the cable and dead men from the crater floor. He intends to use them to secure the buildings in the new camp against the winds. The dilapidated remains of the horse whim can still be found on the southeast corner of the rim. The long beam the horse had pushed still resting on the ground as if awaiting another animal to be attached.

On September 18, 1920, Holland reported that more than half the muckers quit again on payday the 16th. Because of this lack of labor and because of the rain he had to keep all the men on the rig site preparations. He writes that it has rained so heavily that the land between the crater and Winslow is a sheet of water. He put his weekly letter on the train at Canyon Diablo because he could not get to Winslow. But this heavy rain failed to supply their reservoirs. The level of water in their dams did not rise a single inch. At Mr. Hart's dam at Sunshine about 7 miles from the crater, the ground was hardly moist.

This writer has experienced these rainstorms in northern Arizona. In the afternoon the storms will form and move across an area. There may be heavy rains in one spot and just a mile or less away nothing. Staying and risking being struck by lightning and enjoying the coolness the rains bring or leaving is the daily decision when meteorite hunting in the monsoon season of northern Arizona. But these rains though a nearly daily occurrence are brief, and the ground is dry again very soon. Less often does a real storm move through dropping significant amounts of precipitation. That was what happened at Meteor Crater in 1920-21. There was often rain in the area around them, but it never produced enough water to run into the streams and get caught by the dams.

The following week around September 25th the pump was lowered on its concrete foundation. The intake on the pump was increased from 2 inches to 3 inches. The pipeline was finally complete, even though a section was still lying on the surface next to the old ditch. That ground was mostly rock and the trench needed to be deepened. To avoid hiring more of the "very unsatisfactory labor" Holland got a powerful truck and a "rooter" for the digging work. He needed to finish up the pipeline. He had sent for the rig builders to come. They were delayed but would arrive soon. At this point, nothing could remain unfinished that would prevent them from starting to drill.

Holland writes a six-page summary of all the preparation work done up to October 15, 1920. The water system, pipeline, and the slide are featured along with all the other aspects of the work. We learn from this summary that the one excellent workman named Terrell was obtained from the local cattle rancher Mr. Hart. Regarding the casual labor which has been such a problem, Mr. Holland takes off the gloves in this summary report, and calls the men "mostly short stayers and poor workers, aptly described as poor white trash." Such phases would not be seen in a supervisor's report today, but in 1920 a great many things could be said as part of normal conversation. Or written to your aristocratic bosses back in Boston. On occasion, Mr. Holland makes comments and slurs that would make a decent person today blush.

Holland also reports in the summary that new parts were ordered for the pump several weeks earlier. The pumpman who ran it for Mr. Hart was secured. He at least would be familiar with the conditions. The lower dam near the pumping station was raised five feet in height with brush and rock. The leaks that occurred in August during the only good rain were "stopped with scores of sacks of dobe mud laid brick fashion."

This is the view from the interior of Canyon Diablo of one of the canyon walls. There are spots where it is narrow and a dam could be constructed to block the water. In other places, the Canyon is too wide for the type of earthen dam described in Holland's reports. The thick green vegetation in the canyon floor but not seen elsewhere suggests that what happened in Holland's time continues. Even when the canyon appears dry, there is still water just below the canyon floor.

Holland adds some details in the October summary about the upper dam. It was in poor condition, and the water level was very low. No repair work was done to it. "Pigs, sheep, and cattle pollute the water at all times." he writes. There were no means available for them to obtain water from the upper dam storage other than to cut the dam and let the water run to the lower dam. That would clearly contaminate the water in the lower reservoir that was used by the camp. They renewed the abandoned fencing down at the lower dam to prevent the animals from getting at that portion of the water system at least.

In wages and materials, the water system at Meteor Crater cost $6,555.39 as of October 15, 1920. For a little perspective, the pipeline work had required several casual laborers for the better part of several months. It also required the purchase of 10,000 feet of new pipe. Despite all his criticism of the casual workers, the truth is they were paid very little money for their work. The San Francisco office of United States Smelting Refining and Mining Co. paid $2,646.56 for materials for the water system. That left $3908.83 for the labor of a team of men to haul, distribute, and connect pipe, dig the ditches, set up the 13,000-gallon tank after hauling it from the north side of Meteor Crater, building the concrete foundation for the pump, putting the pump on the foundation and later burying the 15,000 feet of pipeline. Holland seems to have tried to hire and keep about 12 men working as muckers and diggers during this preparation phase. That is a great many man hours of labor paid with just $3,900 or less. There was probably a cost for using the more powerful truck and the "rooter" which was not itemized. The food and other camp supplies were also not separately reported.

Two weeks after his summary report the drilling would begin. The water level would not increase at all in the reservoirs during the remainder of Holland's stay. By February 1921 he was much concerned that there would be only sufficient water for the needs of the camp, not the drilling. During the shutdown to dug the tunnel they could have stored up a little water, but no rain fell that benefitted their supply. However, storms were often in the general area around the crater during March and April. On April 15th he states that the water in Canyon Diablo is "very low and muddy and there is still enough for the camp, churn drill could not be supplied for many days."

In the two weeks following his summary of October 15th twelve men were hired to cover a last section of the pipeline. Likely the 5,000-foot portion of pipe salvaged from the crater. Within four days of work, eight of these men had quit. Four of them returned by the next week. Holland remarks that they probably returned "because they had spent the few dollars they had received during the previous week." The ditch was filled in entirely by his report on November 6, 1920. He is satisfied that the pipeline is in good shape for the winter. The casual laborers were all laid off except for two men repairing the dilapidated pump house. Just the regular crew remained on the payroll.

The last mention of the wooden slide which allowed them to pull the salvaged pipe out of the crater is on April 4, 1921. Holland writes that it "has gone down, but fortunately, the occasion for its use has passed." The wood for the timbering of the tunnel was being moved to the tunnel entrance by mules borrowed from Mr. Hart. Most of the wood at the crater was wood taken from the old buildings. The slide itself was made from wood originally in the old buildings. Lumber in the area was very expensive at the time. This was mentioned repeatedly, and figures were given for board foot costs. There was a thriving lumber business going on in the forests near Flagstaff just thirty odd miles away. But there was no benefit to the crater in lumber price. It is possible that all the lumber was being shipped to places far from the mills to support construction in the rest of America.

There was just a little rain on April 4th followed by terrific winds. On April 18th Holland tells Boston that there is no water within 200 feet of the dam. They have been able to keep the tanks at the crater full by digging a hole in the mud at the pump intake. So water was still in Canyon Diablo below the surface. He tells his boss that drilling can not resume until there is water in the reservoir. Two weeks later Holland will be gone, and all this will be the next superintendant's problem.

The Rig

Steam... I am used to steam. I know steam, I like steam power plants. They are reliable, practically indestructible. You fill the boiler with water, burn something in the fire box, oil the moving parts, and they make power. The last few years I have seen more advertisements for gasoline engines in the mining journals. Now they have given me a gasoline engine here at Meteor Crater to power the drill rig. Maybe it was the water worries. We have not had a drop of water added to the reservoirs in months, and the water levels are low. We might have gotten in trouble with steam. Hauling and buying water for a steam engine would cost more money. We would have to get a tanker truck. If we could not get the truck over the bad roads, we would be stopped.

This temperamental gas engine will hardly start sometimes. Especially mornings when it is cold. We set it up the way the instructions in the parts catalog say, but it takes forever to get it to catch. The Union Tool people say we will get used to it and figure out exactly what it needs, but so far we have not, and I resent their simplistic answers. I hope they have better solutions if something really serious happens. They could not give me any help to relieve the way it guzzles gas. We have been using 3 barrels of gasoline a day. I would let it run all the time if it weren't for the amount of gas it uses, then at least I would not have the starting it up troubles. But to let it run for hours while we are dressing tools and doing other tasks is a waste of fuel. I look at it while it is running and a shaft does not look right. I think it is untrue. We can not try to straighten it here, but I think it's the problem with the fuel adjustments. That expert from Flagstaff came out and did very little, but between his work and ours, the machine is better than at first. But I have to get the gasoline consumption down. That would be about half what it is using now. We are hauling barrels from Sunshine every few days in the Ford truck. I am ordering too often and paying an extra charge to have the gasoline forwarded from Holbrook to Sunshine.

Towering above the rim of the crater was the wooden derrick. Steam engines had been the standard power for driving drilling rigs until just about the time of the drilling on Meteor Crater's rim. The isolated location of the crater made a gasoline engine a viable choice. A steam engine could have been fueled by wood, coal or fuel oil if there was enough water for the boiler. But gasoline was coming into its own and was the powerplant chosen. By the middle of September 1920, the site for the drill rig had been cleared, and the rig builders had been telegrammed to come to the crater. The wood and other materials had been on the site for some time. The riggers were delayed for ten days, but they did arrive on Monday the 27th. They began assembling the rig on the next day. The rigging crew was four men who would be assisted by the rest of the men already working at the site. The rig was expected to be completed by the 10th of October. The Republic Supply Company was supposed to send the required materials for two rigs. However, the riggers determined that there was a considerable shortage of both wood and hardware. The required materials were stolen from the second rig to complete the first. By the time the first rig was built the pile of timbers for the second rig was looking skimpy and was short a great many bolts also.

For more than two weeks dynamite promised in both Flagstaff and Winslow had not arrived. The deep cellar required for handling the casing had to be blasted out under the rig floor with black powder instead. It was the only explosive available. Other things too did not go as well as hoped, but finally, on the 15th of October the rig was nearly complete, and the drilling crew was expected to depart California for the crater.

On the first day of November 1920, the drillers began working in three eight hour shifts or towers as they called them. By November 3rd they were already at a depth of 97 feet. The limestone formation was showing many crevices, and it was very difficult to keep the hole going straight.

The drill rig on the rim of Meteor Crater was a cable drill, and there was no real turning of a drill bit. The drilling activity in this type of rig is achieved by raising the heavy drill tools and dropping them on the rocks to pulverize them. The bit on the end of the string of tools is a heavy hollow cylinder of steel with a chisel edge. Behind the bit on the drill string are long sections of rod and devices named "jars" and the cable connector. The jars are connected links of iron that slide within each other and allow the bit to be jerked upward from the rock sharply after each downward blow preventing the bit from becoming stuck. Crater Mining Company received a quote for materials from Republic Supply Company for jars with an 8-inch stroke. Jars came in other lengths of stroke up to as much as 13 inches. A drill tool string might be as long as 50 feet. At Meteor Crater different lengths depending on what parts were broken, off for repair or even just what was available were used. Not always good choices could be made. But a tool string of 30 feet was standard and used whenever possible. At the worst of the troubles, they broke four stems in a week. While those were in California being repaired other combinations of stems and bits and jars were used. Usually, these were far shorter and lighter. Being lighter made them much less efficient.

The hole drilled into the south rim was not a small diameter hole as those into the crater floor had been. The south rim hole began as a 10-inch diameter hole which could be telescoped downward in diameter if necessary. There was 8 ½ inch casing and tools on hand as well as even smaller diameter casing and tools. The drill strings of such rigs could weigh two or three thousand pounds. There was a tremendous amount of force hitting the rocks everytime the bit dropped.

The following photograph is a page from the 1914 Fairbanks Morse Catalogue a company that supplied equipment for drilling. This is likely a type of bit that was present at Meteor Crater for the cable drilling rig that was employed there. The 10-inch auger bits listed are in the range of 600 pounds. The flared cutting edge would wear down. The tooldresser would pound it out at the forge and shape the cutting surface. Circular gauges were used to guarantee the diameter was correct before putting it back down the hole.

Drilling Tools

FOR CABLE SYSTEM

ALL STEEL DRILLING BITS

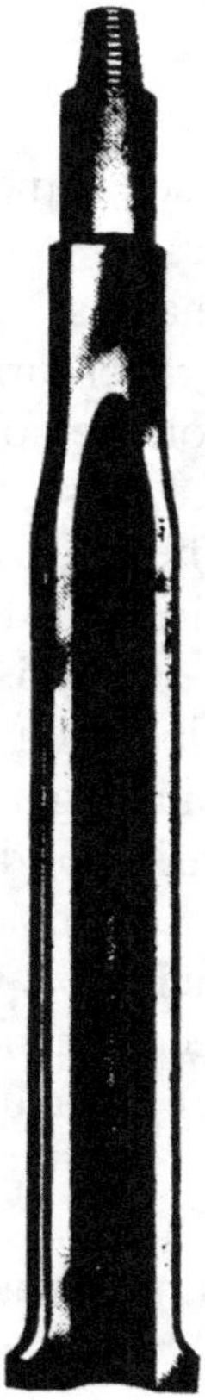

Fig. 90

Size Hole, Inches	L'gth Feet	Size Pin, Inches	Thr'ds Per Inch	Wt., Lbs.	Size Hole, Inches	L'gth Feet	Size Pin, Inches	Thr'ds Per Inch	Wt., Lbs.
4½	5	1⅝x2½	8	135	8¼	5½	3 x4	7	460
4½	5½	1⅝x2½	8	150	8¼	6	3 x4	7	500
4½	6	1⅝x2½	8	165	8¼	6½	3 x4	7	540
4½	5	1¾x2¾	8	145	9⅝	5½	2 x3	7	505
4½	5½	1¾x2¾	8	160	9⅝	6	2 x3	7	560
4½	6	1¾x2¾	8	175	9⅝	5½	2¼x3¼	7	515
5	5	2 x3	7	190	9⅝	6	2¼x3¼	7	570
5	5½	2 x3	7	210	9⅝	5½	2½x3½	7	530
5	6	2 x3	7	230	9⅝	6	2½x3½	7	585
5	5	2¼x3¼	7	200	9⅝	5½	2¾x3¾	7	545
5	5½	2¼x3¼	7	220	9⅝	6	2¾x3¾	7	600
5	6	2¼x3¼	7	240	9⅝	5½	3 x4	7	570
5⅝	5	2 x3	7	220	9⅝	6	3 x4	7	625
5⅝	5½	2 x3	7	245	9⅝	6½	3 x4	7	675
5⅝	6	2 x3	7	265	10	5½	2 x3	7	535
5⅝	5	2¼x3¼	7	230	10	6	2 x3	7	585
5⅝	5½	2¼x3¼	7	255	10	5½	2¼x3¼	7	545
5⅝	6	2¼x3¼	7	275	10	6	2¼x3¼	7	595
6	5	2 x3	7	250	10	5½	2½x3½	7	560
6	5½	2 x3	7	275	10	6	2½x3½	7	610
6	6	2 x3	7	300	10	5½	2¾x3¾	7	575
6	5	2¼x3¼	7	260	10	6	2¾x3¾	7	625
6	5½	2¼x3¼	7	285	10	5½	3 x4	7	600
6	6	2¼x3¼	7	310	10	6	3 x4	7	650
6⅝	5	2 x3	7	270	10	6½	3 x4	7	700
6⅝	5½	2 x3	7	300	11⅝	5½	2½x3½	7	645
6⅝	6	2 x3	7	325	11⅝	6	2½x3½	7	710
6⅝	5	2¼x3¼	7	280	11⅝	5½	2¾x3¾	7	660
6⅝	5½	2¼x3¼	7	310	11⅝	6	2¾x3¾	7	725
6⅝	6	2¼x3¼	7	335	11⅝	6½	2¾x3¾	7	785
6⅝	5	2½x3½	7	295	11⅝	5½	3 x4	7	685
6⅝	5½	2½x3½	7	325	11⅝	6	3 x4	7	750
6⅝	6	2½x3½	7	350	11⅝	6½	3 x4	7	810
7⅝	5½	2 x3	7	375	11⅝	5½	3¼x4¼	7	725
7⅝	6	2 x3	7	405	11⅝	6	3¼x4¼	7	790
7⅝	5½	2¼x3¼	7	385	11⅝	6½	3¼x4¼	7	850
7⅝	6	2¼x3¼	7	415	12½	5½	2¾x3¾	7	775
7⅝	5½	2½x3½	7	400	12½	6	2¾x3¾	7	850
7⅝	6	2½x3½	7	430	12½	6½	2¾x3¾	7	925
7⅝	5½	2¾x3¾	7	415	12½	5½	3 x4	7	800
7⅝	6	2¾x3¾	7	445	12½	6	3 x4	7	875
7⅝	5½	3 x4	7	440	12½	6½	3 x4	7	950
7⅝	6	3 x4	7	470	12½	5½	3¼x4¼	7	840
8¼	5½	2 x3	7	395	12½	6	3¼x4¼	7	915
8¼	6	2 x3	7	435	12½	6½	3¼x4¼	7	990
8¼	5½	2¼x3¼	7	405	12½	5½	3½x4½	7	850
8¼	6	2¼x3¼	7	445	12½	6	3½x4½	7	925
8¼	5½	2½x3½	7	420	12½	6½	3½x4½	7	1000
8¼	6	2½x3½	7	510	12½	5½	4 x5	7	900
8¼	5½	2¾x3¾	7	435	12½	6	4 x5	7	975
8¼	6	2¾x3¾	7	475	12½	6½	4 x5	7	1050

When ordering, state size of pin; style of taper; number of threads to the inch (flat or sharp); size of wrench square; also diameter of the hole in which tool is to be used.

ALL STEEL DRILLING BITS SOLD BY WEIGHT.

PRICES UPON APPLICATION.

The bit would pound the rock hour after hour, being raised out of the hole when necessary to remove the rock that had been pulverized. This was done with a different tool on a separate cable lowered into the hole. The hole needed to hold water at the bottom for many feet. The bailer as it is called would be lowered into the hole and fill with the rock mud churned up by the bit. There was a valve on the bottom of the long cylindrical bailer which would open when it hit the bottom of the hole. The mud would flow into the bailer. When the bailer was lifted, the valve would close. The contents would be raised and dumped at the surface. This bailing would be repeated several times, and then the drill string would be reinserted into the hole, and the pounding would continue. More water would be added as required. Drillers could collect samples of the contents of a bailer for testing later. This collecting was done at Meteor Crater. Holland would allow the contents from the bailer to drain and then he would pan the concentrate gathering any metallic matter with a magnet. Dimethylglyoxime, and the other reagents for nickel testing were at the crater. He and Mr. Plumb after him were constantly trying to determine if the material in the hole was meteoritic or just pieces of casing or chunks of lost tools and drilling bits.

This method of drilling goes back a couple of thousand years and is still used some places today. It is less expensive than rotary drilling and requires much less complicated machinery. It can make steady progress through rock. But it has its limitations. One being that the hole must hold water so that the dust and gravel formed by the pounding can be removed by the bailer. Otherwise, the bit pounds on the powder it has already made and does not hit the rock to continue cutting downward. Also, this type of drilling has great difficulty with badly fissured rock formations. The bit strikes the edge of the sloping crevice and slides away at an angle along the crevice. The bit can not create a shoulder to pound on. The long string of rigid steel above the bit is not supposed to bend. If the bit wanders or glances as it is called down a crevice, then something in the drill string can easily break. At Meteor Crater they began using a 10-inch drill bit. As problems increased, they got down in size to a six-inch string of drill tools. The six-inch tools were usually run down the outside of the main hole to try and sidetrack lost equipment or aid in straightening a crooked hole. The six-inch tools were on another cable called the sand line. Near the end of Holland's stay, he recommended to Mr. Jennings that they restart after the hole was cleared with a 12 ½ inch bit and tools. Jennings approved this, and the larger tools were ordered.

All of the stem breakages in the first few weeks occurred at the same spot in the tool string. By December 30, 1920, all four of their drill stems and sinkers had broken on the long stem part right at the box near the end. As of that same date, there were an underreamer and two bits lost in the bottom of the hole. They would eventually get long-awaited correct fishing tools, employing them they would quickly extract most of the lost parts. The 8¼ inch drill bit would be recovered on February 4, 1921, but it was broken off leaving about a foot remaining at the bottom of the hole. The third lost bit was also recovered on the same day. The underreamer, however, was not recovered then or ever by fishing it from the hole. The large chunk of the 8¼ inch bit moved around the bottom of the hole from time to time and was finally sidetracked and bypassed.

The dryness of the hole and the large crevices worked together to add to this disaster of a drilling program. The crevices in the rocks let the water poured in at the top run away out the bottom. So they suddenly could be working in a dry hole. If the pulverized material from the drill also ran away down into the crevice, then the drill could keep cutting down crooked until something snapped. Sometimes the fissures were filled with sand which ran into the drill hole preventing any progress by the bit. Without water in the hole, the drillers could not use the bailer. The running sand could get behind the bit and other tool string parts locking them down in the hole so it was difficult to pull them up. The same might happen with small boulders that caved off the sides of the hole. Many times a leaking hole would hold water long enough to make some progress. But they would have to add water from the surface frequently. They could also send down several trips of a water-filled bailer. It would dump water when it hit bottom. Then the drillers could agitate the rock powder into mud that could be raised by the bailer to the surface before the water leaked completely away.

Sometimes drillers in the 1920s would try to seal up fissures that were encountered in a hole. They would dump debris and clay and even cement into a hole attempting to create a surface that the bit could work. Despite later remarks by Barringer to the contrary, this was a very ingenious drilling crew with much experience drilling in fissured limestone, according to Holland. That would lay much of the blame for their problems on the constantly breaking drill stems.

If a hole became crooked drillers would redrill the hole over and over until it was straight. Often at Meteor Crater new fissures encountered would prevent the hole from remaining straight for very long.

Meteor Crater formed when an asteroid hit the ground at over 25,000 miles per hour shattering the rock under the rim. The drill bit encountered fissure after fissure. Some of the crevices were so large that light from lanterns sent down on ropes would disappear completely into the cavernous cracks. The string of drill tools repeatedly broke while Holland and his crew fought their way through the limestone. There were hopes that the sandstone would be easier to drill since it was softer. The drill bit might have a better chance to create a shoulder to pound on and continue drilling straight. Holland and the crew got just through the bottom of the limestone. At 290 feet Holland writes "the washed sludge was practically all brown and yellow saccharoidal sand with a great deal of iron and steel. The latter gave no reaction for nickel" The last mention of the contents of the bailer was from a depth of 295 feet, and it was sand. There was just a little of the Moenkopi Sandstone left after the drill site was cleared and leveled. The Kaibab Limestone at Meteor Crater was known to be approximately 265 feet thick. At their final depth of 312 feet, they did just reach the top of the Coconino Sandstone which is 700 to 800 feet thick.

One of the most valuable men working on a drill rig was the tool dresser. In 1920 he was paid about 80% of what the drillers were paid and much more than the muckers and roustabouts. After a time the driller would decide that the bit was worn flat and he would pull it from the hole. Or it could be inspected during the time that the hole was being bailed. It was the tool dresser's job to place the heavy bit in the forge and reshape the cutting surface and resize it to the correct diameter. The bits were often several feet long, and some were fluted on the sides for water circulation. The tool dresser would have to harden and temper the metal after reshaping it. The driller would then reattach it to the end of the string of tools and return to pounding the rock. While the dresser was working on the bit, the hole might be bailed or the equipment maintained and adjusted. If there was a spare bit, it might be attached and used while the other was reworked. Bricks likely from the forge are still lying around the drill site on the crater rim close to where they were when the site was abandoned.

The 1920's were a time of change in the drilling industry. As mentioned already there was a shifting from steam to gasoline power. Rotary drilling was taking the place of churn drilling. There was also the introduction of steel cable for holding the drill tools and raising them to pound the rock. The thick manila rope was the standard for cable drilling before the 1920's, and it was still in wide use especially for holes of only 200 feet or less such as water wells. But as drills went deeper, steel ropes became the cable of choice. Each drilling rig had several long lengths of cable. One was on the drill tool string; another was on the bailer, a third might be a sand line. Others were used to hold equipment down. At Meteor Crater much strong rope was used to hold the rig in place against the tremendous winds that ravage the area frequently. The rig actually had 12 guyed cables connected to it.

Quotes given for parts show prices of cable sockets for both manila and wire rope. So both types of rope were likely used at Meteor Crater. 2½ inch rope is shown on an inventory of unused parts in stock at the crater. A quote was gotten for a 2½ inch New Era Manila Rope Socket from the Republic Supply Company.

Manila drill rope came in several diameters. The 2½ inch rope came in lengths up to 4,760 feet. At 2.47 pounds per foot, it would have been heavy even in the short length used at Meteor Crater. It had a breaking strain point of 53,665 pounds but was recommended for best results to be used at only 5% of breaking strain. That amount would be just 2,683 pounds of load. That was right in the weight range put on it at Meteor Crater with 30-foot stems. The 1914 Fairbanks, Morse and Co. catalog shows a selection of wire cables even years earlier then the Meteor Crater drilling. Wire cables were usually six strands of sixteen wires each with a hemp core for flexibility. The strength of the wire rope was far greater than the manila and estimated for the thinnest cable at 14 tons and the thickest at 53 tons. Nothing is said about the breakage of wire cable due to metal fatigue after months of continuous use. Metal fatigue would certainly be a consideration tested for today.

Drill derricks were a common sight in the last century as petroleum demand increased. In places, they covered the land like a forest. At the top of the derrick of a cable drilling rig was at least one large pulley which carried the rope from the walking beam to the drill string. As the walking beam moved up and down the drill string moved up and down, and the drill bit and long string of tools would repeatedly strike the rock. On the rig at Meteor Crater, the shaft of the bull wheel was made of wood. The "pitman" was a connecting rod hooked to the bull wheel at a point near the wheel rim. The other end of the pitman was connected to the rear end of the walking beam. This eccentric connection to the bull wheel created the up and down motion of the walking beam. The pitman on the rig at Meteor Crater was also made of wood. Since both the bull wheel and the pitman were under load and in continuous use, Mr. Holland had spares assembled out of the materials supplied for the second rig in case the parts broke down. It is hard to conceive that such a large and heavy piece of equipment like the drill rig would be shipped with these parts made of pine wood instead of steel.

Derricks came in a variety of heights. Some were short for shallow water well drilling while others were much taller for deep drilling such as needed for extracting petroleum. Often illustrations of the derricks will show a square platform like structure in the upper half of the derrick. Such a platform was on the derrick at Meteor Crater. Connecting stems and cables needed to be done at the height of their upper end. The cellar at Meteor Crater was eighteen feet deep, and the preferred length of stem was 30 feet. When the rod needed to be detached or serviced its upper end would be somewhere around twenty feet in the air if you account for a drill bit of 6 feet which would have to be raised out of the hole to rest on the cellar bottom. The photograph of the south wall of Meteor Crater showing the derrick taken by Holland shows a building surrounding the rig and the platform structure higher up the rig.

The gasoline engine was mounted on a concrete foundation and connected to the rig by a belt which ran on large wide wheels. The engine was almost certainly made by the Ideal Gas Engine Company. The name "Ideal" is repeatedly used in correspondences to and from Holland. It was difficult to start, especially on the cold morning. For a long time, it was consuming about twice the gasoline intended. After several repairmen and the drilling crew worked on it the engine was using only 80 gallons of gasoline per day. It ran more or less continuously. When things were working right, the drilling was going with three shifts of eight hours. Detailed instructions on how to start the engine were finally mailed from Union Tool Company, Torrance, California to the crater in response to a wire Mr. Holland sent on the third day of drilling November 4, 1920. The instructions given are as follows:

"(1) Be sure that the engine cylinder is clean and free from water and is not flooded with gasoline. Either of the above liquids can be removed from the cylinder by taking out the inlet valve cage and inserting a mop cloth through this opening. (2) Test spark to see if same is O. K. at the spark plug. (3) See if the gasoline is flowing freely to valve "G" as indicated on attached sheet No. 26. (4) See that the gasoline pump is working satisfactorily so as to supply gasoline to carburetor body (1205) page 26. (5) Be sure that water valve, marked "W", page 26, is fully closed. (6) Retard spark by shifting part 1407, page 38. (7) Open relief cock "A", page 4. (8) Fill a small squirt can with gasoline, move the crankshaft to extreme outer position, and give three or four squirts of gasoline into relief cock "A". (9) Insert squirt can into air pipe 1212, page 26, and give three or four squirts of gasoline. (10) Grasp flywheel and turn engine as rapidly as possible. Open gasoline valve slightly at the same time. Should engine not start on first two or three turns, squirt a little more gasoline into pipe 1212, opening gasoline valve slightly at the same time. Under ordinary conditions the engine should start, and by properly regulating the gasoline supply with gasoline valve "G" a proper running of the engine should be obtained. A little practice will soon enable the operator to know exactly how many turns to give to the gasoline valve for starting and running, in which case it would not be necessary in starting to squirt as much gasoline into pipe 1212 as hereinbefore mentioned."

Those instructions are worse than the ones that come nowadays with furniture that has to be assembled. I can visualize the men who are familiar with steam standing on the rim of Meteor Crater with the sheets of drawings and the instructions trying to find the various parts and set them as described. I can almost hear someone yell "contact" as one or two other men spin the big flywheel as fast as they can by hand. Mr. Clark of Union Tool wrote in his instructions letter that they would figure out the proper settings and procedures with time and practice. A first indication of the really poor customer service that was to come from Union Tool and later National Supply Co. which acquired Union Tool in December of 1920. In just a few days Holland would find that the drill rods and tools were very poor that they had received from Republic Supply as well. He would complain but never get any satisfaction from these manufacturers and suppliers. Mr. Holland would finally be forced by this bad service to search for other providers to get relief from the equipment troubles plaguing him.

It's clear from Holland's reports that the gasoline engine would not have been his first choice. However, a letter from Mr. Jennings to Superintendent Holland about the engine contains wording that is as strong as any ever written to him. He is to make the gasoline engine work. The manufacturer has said it is the best choice for the isolated crater location. No other choice of a power plant is going to be made available to Holland for use at the crater. So Superintendent Holland fights the battle of the gas engine, and eventually, the machine is working acceptably. But when it does break down it begins the sequence of disasters that are never really overcome before Mr. Holland leaves.

One interesting feature of the gasoline engine is that it was equipped to inject water into the cylinder to be mixed with the gasoline. The bottom portion of the carburetor had water inlet and outlet pipes and jets to squirt a mist of water into the engine intake to mix with the gasoline mist. This was an early practice to combat knocking caused by the too rapid detonation of the gasoline. The fuel-air mixture with added water would burn smoother, and the engine would not bang as much. The water vapor also provided some cooling to the cylinder. It was very important that the water was turned off by closing that "W" valve before stopping the engine. It was slightly opened only after the engine was started. Water in the engine would not only prevent it from starting but would rust the inside of the cylinder if the engine was stopped while water was being added to the fuel-air mix. The engine would be run for a short time without the water to purge it from the cylinder. Then the engine could be shut off.

The drilling began on November 1, 1920, with the engine running 24 hours a day. They must have stopped the engine to do other activities on the rig because he comments in his weekly report on November 13th that, "the "Ideal" gasoline engine is very hard to start, especially when cold, and its consumption of gasoline is very high. I have taken it up with the manufacturer, in hope of getting some relief." The "Ideal" in that quote is both sarcastic and one of the places where the manufacturer is identified. On November 19 the engine broke down with the failure of two essential bearing studs that could not be replaced locally. Holland writes they had to be ordered from Los Angeles. But the likelihood is they came from Union Tool in Torrance, California a few miles south of Los Angeles.

The replacement parts were supposed to arrive on November 21 a Sunday. However, they did not arrive until Wednesday the 24th. It was just an extra three days but turned into a fateful delay. During the four days which the rig was idle, the casing became cemented hard in the hole. It could not be pulled out. This was likely because the powdered limestone formed a cement-like substance around the casing pipe. It took just an hour to install the replacement parts and get the engine running again. But the damage was done. Over the next ten or eleven days, the drillers tried everything they could to extract the casing. They even blasted the hole with torpedoes. Torpedoes are cylinder-shaped devices that held explosives and could be run down the hole. The shock of the detonation was supposed to break loose any rock holding the casing pipe. Finally, near the first days of December, the drillers shot off the casing shoe and late on December 4th blasted off the bottom 14 feet of the casing with explosives. They had to drill out that fourteen feet of steel casing and then redrill the hole. Pounding down and breaking up fourteen feet of steel pipe was of course very damaging to the drill bit.

The plan before the casing became cemented in place was to change from the 10-inch tool string to the 8¼ inch drill tools then redrill the hole. That is what they did. One instance that shows the quality and experience of Holland was that he had even at this point seen the possibility of equipment breakage in the fissured rocks. He had ordered spare jars from Republic Supply Co. on November 27 while they were struggling with the cemented casing. On December 6th the original jars of the 8¼ inch tool string did break almost immediately after they changed from the 10-inch tools. Holland ordered a second set of replacements by wire. The first spare set was said to have been shipped on December 1 about when they shot off the casing shoe. The second set he was told shipped express on December 7. They met all the trains at Canyon Diablo Station, and no jars had arrived by his weekly report of December 11. A set finally arrived on Sunday, December 12th and was connected to the drill string. Holland had ordered the heavier type of jars that were commonly used in California. They were much stronger than the ones that came as original equipment with the rig. During the night shift of December 13, the casing went down easy to the bottom of the hole at 200 feet. By the following morning, the hole was at 218 feet. Pieces of the drilled out casing were still coming up in the bailer then as they would for a long time.

On December 4th Holland reported that the engine had finally been adjusted by them and the gasoline consumption was down. On a full day of running it consumed 80 gallons. That was close to the 72 gallons per day the manufacturer had said. For the entire month of November, the engine had been consuming 3 barrels of gasoline per day. With a barrel being 42 gallons the consumption had been 175% of what it should have been. The repairs and adjustments made at the crater were based on the belief that a shaft on the engine was not true. In characteristically poor service the recommendations of Union Tool Company provided no relief to the problem. It was again the ingenuity of the drill crew and Holland that solved the problem.

Drill rigs were large, and by today's standards, dangerous pieces of equipment with little worker safety engineered into them. The walking beam on a drill rig was a moving timber part often 25 feet or more in length. The spinning wheels were uncovered though often enclosed in a building. Workers needed to keep their wits about them to not have an accident. Accidents did happen, and men also got sick. Meteor Crater is an isolated location today. In 1920 it was even worse since there were no telephones at the crater. Everything for the rig and all their supplies had to come by a freight train. Anyone seriously hurt, or sick was also sent somewhere by train for treatment. The medical facilities in Winslow were limited.

The cost of the rigs and of preparing the drill site as of October 15, 1920, totaled $34,576.38. Beyond that, it cost more than $10,000 for the camp buildings and the mess house. The water system to supply water for human use and the rig cost $6,555.39. The Overland automobile, the Ford truck, and other general expenses came to $3,907.91. The total expenses of Crater Mining Company as of October 15, 1920, were $62,919. Not one foot of this first hole had yet been drilled. That was still two weeks away. United States Smelting Refining and Mining Co. had committed their subsidiary Crater Mining Company to spend $75,000 on the entire project of up to ten holes. There was little of that money left when drilling actually commenced on November 1, 1920. By the end, they would spend much more than originally agreed.

The Magnetic Survey

I telephoned him from Winslow yesterday and from Canyon Diablo today. I did not reach him. I guess the lines are down. They are bringing him back again. I am not envious. That's not really how I feel. It is just that he is making so much money and I question the wisdom of the expense. They did a magnetic survey, and now they want another done right away. This one is going to take forever. We are weeks away from beginning to drill, and he will still be working on this survey when we are spudding in. That is if we do the survey their way. I don't think the men in Boston understand how big the crater is. To survey it with magnetic readings the way Moore and Barringer ask is going to take months, but they want it done quickly. I don't know how to reconcile those two things or pay the bill. We have already gone through much of the money, and I am still far from having the drill site prepared or the pipeline finished. The contractors are still building the camp, and it is costing far too much, but that is how it is here. Wood is very expensive and labor is in short supply and of poor quality. They want $80 per thousand board feet for rough sawn lumber and the men sometimes are union miners who want too much. I did not need to add this survey to my work. But if it will help to find the best location for the second drill rig that would be useful. The first site was chosen because something threw the rocks up much higher into an arch and it is hard to figure out what besides the asteroid could have filled that displacement. Barringer is convinced about the south rim, and he has sold it to the men in Boston for the first rig site. I just do what they tell me. So back to Canyon Diablo by the terrible road tomorrow to try and get Fay on the telephone. I have no doubt he will hire on for such a big magnetic survey. I'll give him Moore's sketch and Barringer's instructions and let him figure out how to please them both; it's beyond me how he will do it.

Explorers in any arena of investigation want to use all the technology available to them. The same was the case at Meteor Crater in 1920. The first investigation by Grove Karl Gilbert in 1891 had relied heavily on negative results of a magnetic survey to declare the crater was volcanic in origin. Barringer strongly believed that the massive iron asteroid could be detected by studying the magnetic field of the crater. By today's standards, there was not much equipment available. They had no gravity meters, magnetometers, ground penetrating radar, or metal detectors. They had no enhanced GPS as a surveying aid. They had a surveyor's compass or its cousin the dip needle from which to choose. These instruments had not changed much in a couple of hundred years. The compass is familiar to everyone. It has a magnetized needle that points toward the Earth's magnetic pole. It does not point toward true north usually and if it does it is by coincidence. The angular difference in the direction which the compass needle points to, from that of true north, is called magnetic declination. It can vary widely with geographic location. A somewhat less familiar characteristic of the Earth's magnetic field is that it has a local dip everywhere. The magnetic lines of force which surround us have an angle to the surface of the Earth. This angle can be measured using a dip needle also called a dip circle. It is essentially a compass turned on its edge. The magnetic needle will align downward at the angle of the local magnetic field and this angle can be measured on a precision scale. Dip needles can be very sensitive to any local magnetic material. Tests were done on Gilbert's dip needle before coming west to the crater. The instrument was supposedly tested at a Naval Yard and was able to detect a pile of iron cannon balls at a great distance. It was considered by him to be sensitive enough to bring to Meteor Crater. Gilbert began very confident he could find a buried iron asteroid large enough to create such a vast hole. But his lack of knowledge about impacts though understandable for the time led him to make the wrong conclusion that it was a volcanic steam explosion which formed Meteor Crater. Barringer would have a magnetic survey conducted as well. He would find in the data several locations where there appeared to be deviations from the normal magnetic field. Why did Gilbert not find any evidence in his magnetic study of Meteor Crater? It may have been that he had his expectations too high. The variations in the magnetic field recorded in these later studies were very small. So small in fact that they were hardly larger than the background fluctuations that occur naturally during the course of a day. Gilbert was likely looking for a large swing in the needle of his instrument caused by the enormous single mass of iron he believed was required to form the crater.

Dip needles could be effective tools for locating buried iron objects if used properly. Small dip needles with a long handle were run close to the ground to find buried manhole covers, pipes, and other iron objects. The magnetic needle would swing down and point to the iron object underground. They were simple metal detectors which only worked with ferrous metals. Barringer and USSR&M Co. would never obtain a dip needle for use at the crater. They would use what Mr. Fay the surveyor had which was a sensitive compass.

Barringer and the officers of United States Smelting Refining and Mining wanted another magnetic survey done at Meteor Crater. They wanted to determine the best locations for drilling large diameter holes to find the buried asteroid. They had Barringer's original magnetic study which showed a location on what was called the MN line. MN would usually refer to magnetic north, and there is little reason to think it does not apply in this case as well. That line ran through a spot marked on the south rim with a flag. It was near the top of the uplifted arch of rocks and the south cliffs. It was Barringer's favorite guess at where the asteroid rested. Mr. Holland was tasked with the responsibility of getting the magnet survey done and of converting the data into maps and diagrams; then having that material printed and distributed to all the individuals involved.

Barringer writes to Holland the first week of July 1920 in a handwritten note that Mr. Fay should be gotten back out to the crater so that Professor Magie can witness the way the measurements are taken for himself. Professor Magie will be one of Barringer's experts in interpreting the final data, and he is going to be at the crater for a visit. Barringer believed when he wrote this letter that the earlier survey was already completed and that for Magie to see the measurements being made would require calling Mr. Fay back. But it appears that Mr. Fay was still at work, but was near the end of the first survey for Barringer. Truthfully this was just the beginning of magnetic surveys and the extensive work of Mr. Fay at Meteor Crater. Early in August the prints of the first magnetic survey, profile drawings of the crater on the MN line showing the new shafts, along with prints of the surface improvements were sent to Messrs. Moore and Anderson in Boston.

Some details of the magnetic survey are a mystery since Mr. Holland included no copy of the survey map in his papers. But since the location of the MN line is known and would become the location of the drill site it is possible to reference some of the survey details to that line and location.

Fay obtained normal readings at most of the places where he took observations. However, there were stations where he read deviations in the local magnetic field. For example, he reports, "variations between Stations 24 and 10 suggests that a body or bodies affecting the needle is between Station 10 and the South MN." The word "needle" is a clear tip-off for us today that he had dip needle or compass equipment and none of the new electrical devices then being devised. This was a time where electricity was coming into its own. Every field of science and medicine had revolutionary devices being invented. Some were made by serious scientists and others by quacks. So-called "doodle bug" devices were being used to locate oil, gas, gold, water and anything else. They were operated for the most part by confidence men. But Mr. Fay we will find out had only a transit compass. Holland further reports Mr. Fay's results in his August 10, 1920, letter to Moore and Anderson with the following. "The variation at Station 8 is west of the normal and would appear freakish, or it may be due to a separate mass somewhere to the north of Station 8. The fact that the variation was found to be normal at a moderate distance around the west side of the crater north of Station 8 might possibly be due to the body being immediately under the observation station." The vague nature of this last passage is what will characterize the study and led to the criticism later. There are no follow up readings taken to determine the size of this "freakish" magnetic anomaly. No measurements are made to determine how the reading slowly fades to a normal reading around the area. Just the statement that the reading is again normal at a moderate distance. The other anomalies recorded will be similarly lightly studied.

Holland did as Barringer requested and got Professor Magie and Mr. Fay together. He reports the reaction of Mr. Magie which he received in a letter from the professor. "On the whole the magnetic variations look right, though the extreme one on the west is off, and if genuine, as I have no right to doubt, would indicate a local mass, in the upper part of the rim, and if this is the case, the same explanation may do for all the variations. Still, the systematic way in which they are distributed indicates that Fay has really got evidence of the big meteoric deposit. I send him my compliments and congratulations. If he is still retained on your work, I would suggest that he work the ground over carefully beyond the western station that shows abnormality. He may get evidence of a big mass that you could reach by shaft on the rim." There is never any pursuit of that anomaly or real detailed study by Fay reported for that area. Certainly, no shaft was sunk there to locate a close to the surface mass. This seems strange after all the dozens of holes and trenches blindly dug for little or no other reason than pure exploration and claim improvement. To have a possible easy to secure mass not investigated seems very strange these many years later. It is possible that until this very writing no one knows of this detail in the magnetic survey of 1920. Modern equipment much more sensitive and reliable might also indicate this magnetic anomaly. An attempt to find the mass might not be a difficult project now.

By August 28th Holland's communications indicate that there are already issues being discussed about the magnetic survey data. Questions are being asked about the variations seen at individual stations over the course of the day. The readings fluctuate at all locations. Mr. Fay has begun taking his measurements at three times during the day at each station and averaging the results. This is the normal and established method used when variations during the day are seen. A hundred years ago there was no clear knowledge of how the magnetic field of the Earth varies during the day. Today we understand that 90% approximately of the field is generated by the internal bipole magnet of the Earth's outer core. But there are daily changes caused by fluctuations in the ionosphere and magnetosphere. These may cause changes which can deflect the surface field by as much as a degree. These changes can be over a time range as short as a millisecond to very long intervals. This could make all of Mr. Fay's compass readings suspect were it not for the general pattern he recorded for which the crater seems to be responsible. Also, he does record a normal reading everywhere at stations away from the crater. He may not have understood the reasons for the daily variations but his averaging the results of several measurements at each station seems to have let him recovered some real data.

Compasses sold today will often come with instructions that advise the user of the dangers of making readings near rocks that may contain magnetite or in or near buildings with equipment that may have iron or magnets. Mr. Fay recorded quite small variations, and in the end, no one could make a definitive interpretation of his data. The survey was such an enormous project that he seems to have gotten caught up in trying to gathering the whole crater's magnetic signature and failed to define any of the actual anomalies he found with fine enough detail.

By the middle of August 1920, Boston has requested the expanded magnetic survey. On August 23 and 24 Holland tried from both Winslow and Canyon Diablo Station to telephone Mr. Fay but the lines were apparently down, and Holland had not gotten a hold of him yet. Holland does reach Mr. Fay who returns for this even more extensive survey "for departures from normal magnetic variations in and around the crater." That phrase might be called the job description and might explain some of the results obtained or not obtained. By September 11 he has been at that work for the past week. In Holland's weekly report of September 18, he notes simply that "Mr. Fay has continued this work and is progressing well with it."

CONSULTING MINING ENGINEER
AND GEOLOGIST

DANIEL MOREAU BARRINGER
1242 REAL ESTATE TRUST BUILDING
PHILADELPHIA

October 20, 1920

Mr. L. F. S. Holland,
    Crater Mining Company,
        Winslow, Arizona

Dear Mr. Holland,

    I hope that Mr. Fay took account of the fact that there are a number of pipes in the central portion of the crater in abandoned drill holes. Before we understood the use of clay in drilling we lost several lines of pipe. The log will doubtless tell just where these are. But as all of the drill holes were in the central portion of the crater, generally speaking, I do not think it possible that these pipes could have affected the southern portion to which, as I understand, Mr. Fay's magnetic survey was principally confined. I think I mentioned the fact to you or to Mr. Fay that there was a lot of pipe stuck and abandoned in the old drill holes but I cannot remember positively. Your maps would show the location of all the drill holes and if Mr. Fay's compass showed marked deviation near any one of them it is not at all improbable that it was due to the presence of the pipe in the hole.

    Mr. Jennings has sent me a copy of your telegram stating that the rig is finished and that the crew has been telegraphed for from California and that you will probably start drilling next week. I wish I could be there to see the start made and shall await with interest copies of your weekly reports regarding the progress which is made.

        Yours very truly,

        Daniel Moreau Barringer

In the report of September 25th, we learn much more. Mr. Fay is still engaged in the survey work. He is attempting to determine the mean declination for the region. This likely means he was working out beyond the crater on the plains. He is taking three observations at all of the important stations at different hours of the day. As of that date, the results were not complete. However, Holland relates that they appear to indicate that the declination is about normal at a distance of 2000 feet from the crater. The greatest variation so far found was in the central shaft at 155 feet down. That was the deepest depth accessible in the shaft. The observations on the crater floor yielded erratic results. Holland remarks that they are being checked carefully. There is by this time a great amount of abandoned pipe and casing in the holes drilled into the crater floor. Barringer will remark about this when he reads the survey results. Maps of his old drilling locations have been furnished to United States Smelting Refining and Mining. The consensus seems to be that deviations caused by left behind buried iron in the crater floor will have little to do with the new operations at the crater. That abandoned metal is thousands of feet away from where they will be drilling. Barringer would be expected to have the best knowledge of what metal had been abandoned in drill holes on the floor. That knowledge is later proved to be lacking. Holland is hopeful that the "isogenic chart" of the magnetic field would be drawn up in the next few days. He hoped that it would "give valuable information." The later criticism of the survey and the description of the map Fay does create, raise questions about whether an isogenic chart showing lines of equal magnetic declination was ever produced. The criticism is in response to the single spots marked with variations without further work to determine the local shape or extent of these anomalies in the magnetic field.

Mr. Fay is still hard at his survey when Holland sends off his October 2, 1920, weekly report. Even after all the work, his survey map revealed several stretches of territory that were lacking observations or had incomplete observations. He was then at the task of filling in those gaps. Holland reports that the fluctuations in readings taken at different times of the day can vary by as much as 10 minutes of a degree. This would confirm what we know today about the effects of the ionosphere and the magnetosphere. He is still using the three observation method and averaging them to get a mean for the declination at each spot. However, some of his variations are little more than the daily fluctuations he is seeing. Possibly another shadow cast on the validity of all the readings.

Holland has never used the words "magnetic inclination" in his reports. Barringer comments in several of his correspondences that he would like to buy, get, or lease a dip needle instrument for Mr. Fay to use. Only his surveyor's compass was employed in the work. The dip needle would have been of great value since it would swing down lower than the local inclination when over an iron mass. Fewer readings might have been needed to locate anomalies with a dip needle than with a compass. However, Gilbert had done a dip needle survey during the many days he was at the crater, and he saw no strong influence from buried iron. He also thought there should be one mass under the center of the crater. He was not looking for tiny variations from many smaller masses underground around the rim or on the slopes. He expected a single solid mass of iron, enormous in size was buried within the range of his dip needle. Gilbert was confident that he would see indications on the instrument if the crater formed from an asteroid impact which had scattered the iron fragments found on the surface. No dip needle variations for Gilbert meant the meteorites found on the plain were coincidental and not from the crater-forming event. Holland and Fay were in a harder position they were looking for tiny magnetic variations with a compass an instrument not ideal for the task. The question remains as to whether a dip needle would have even responded to iron masses buried a thousand feet down. It would seem unlikely since no indications were found by Gilbert's dip needle.

By the time of the magnetic survey in 1920, the iron fragments and the shaleball meteorites had been extensively studied. On September 11, 1920, in a letter carbon copied to Barringer and Professor Magie, Mr. Moore suggested an additional study be done by Mr. Fay or Mr. Holland. On the northeast side of the crater, a great many iron fragments and shaleballs had been discovered in the shallow excavations. He says that magnetic tests should be done on specimens to be uncovered by a new excavation. The officers of United States Smelting Refining and Mining wanted to learn what the polarity of these specimens was when they were still in place in the ground. This could be demonstrated with a compass brought close to the iron meteorites and shaleball meteorites. There were two possibilities expressed in Mr. Moore's correspondence. The first was that the magnetism of the buried meteorites might have been caused by the Earth's magnetism in which case all the fragments would show an alignment to magnetic north. The fragments would have north and south magnetic poles themselves. He felt if this were found to be true it would be evidence of some value. If however the other possibility was found and the magnetic poles of the fragments pointed in all different directions then the fragments were permanent magnets whose magnetism did not come from the Earth's magnetism.

Mr. Moore says that Holland should partial uncover the iron meteorite fragments and shaleballs and prepare a sketch map of their positions and mark with a line their magnetism. Another line should be drawn showing the direction of magnetic north. He suggests that Holland is careful not to test pieces that are adjacent to each other because there may be local influences between the pieces. Six to twelve pieces he thinks is enough of a test sample to show results. This is actually a remarkable study Mr. Moore is proposing. It is a very early investigation into what is today a fully developed area of geological field work. Paleomagnetism is widely used today to determine the magnetism recorded in rocks. It can, for example, be used to distinguish one lava flow from another at a volcanic crater where flows look the same and overlap or pile up. As the lava cools the magnetite that crystallizes will be aligned to the magnetic field of the Earth. Later or earlier flows tested will have magnetic orientations to the magnetic field from the times of their eruptions. Today we know that the magnetic field of the Earth varies in strength over time and that the poles move and the field has even reversed many times in the past. The implications at Meteor Crater for aligned magnetism in buried meteorites could be important.

If thousands of fragments of iron or several huge masses resting close together underground had a single magnetic orientation their various individual weak magnetic forces might add to each other making finding the masses easier. But if the fragments were magnetic before entering the atmosphere and landed with random magnetic

---

F S Holland

PRINCETON UNIVERSITY
PRINCETON N J

W F MAGIE
Dean of the Faculty

September 16, 1920

My dear Mr. Moore:

I received on my return home the copy of the letter which you sent to Mr. Holland dated September 11th, about the magnetic observations which are to be made at the crater.

I think you would like to know that I made observations on the magnetic condition of several of the shale balls found by digging in the northeastern wall. They were all oxidized or covered with a thick shell of oxidized iron and they show local polarity at many points on the surface. There was a general tendency to show more north poles on the lower end and more south poles on the upper end but it was not well marked. I think it would not be worth while to go to the trouble of excavating more of these shale balls for the sake of studying their magnetism. There is enough magnetic oxide formed around them to make the magnetic condition very confused.

I have not heard from Mr. Hansell but am holding tomorrow open for his expected visit.

Yours sincerely,

(Signed) W.F.Magie

To-

Mr. C. F. Moore,
Boston, Mass

orientations then their weak magnetic forces would cancel each other. There was a third possible conclusion not offered by Moore. The fragments might not have any magnetism when in the ground. The specimens tested in laboratories for the last several decades might be magnetic from having large magnets placed on or drawn across them. The meteorites might not have original magnetism at all.

Professor Magie received Mr. Moore's letter dated September 11, on September 16 Magie sent a letter back stating that while he had been at the crater on his visit, he had done magnetic tests on shaleball meteorites found by digging in the northeast wall. He writes further "They were all oxidized or covered with a thick shell of oxidized iron and they show local polarity at many points on the surface. There was a general tendency to show more north poles on the lower end and more south poles on the upper end, but it was not well marked. I think that it would not be worthwhile to go to the trouble of excavating more of these shaleballs for the sake of studying their magnetism. There is enough magnetic oxide formed around them to make the magnetic condition very confusing." Magie's remarks may or may not indicate that the meteoritic material has its own magnetism, just that it responds in a variety of ways to a compass brought near.

It was almost too late by the time of Magie's response to stop the shaleball study. Holland had already acknowledged Moore's request for the study by September 18th stating that "The excavations on the northeast side of the crater for shaleballs and their polarity shall have immediate attention." On September 20 Moore writes Holland and encloses a copy of Dr. Magie's letter and stops the excavation and magnet study of shaleballs. Whether or not the digging was done and the compass readings and drawing were made is unknown. Holland never further mentions the subject. It is possible it was done, and he chose never to bring it up or more likely he had not gotten to it yet, and Moore successfully stopped it.

If they had gotten readings from the meteorites in situ which had formed by the magnetic field of the Earth what would they have seen? Would it be the magnetism imparted to the fragments over the eons that they have been in the ground exposed to the weak field of the earth's magnetism? This seems unlikely. Would it be some kind of magnetism created as the magnetite formed during the conversion to iron shale? The magnetic field of the Earth is changing all the time. It is unpredictable. The position of the north magnetic pole is moving in a northwestern direction. Fifty thousand years since the crater formed is a long time of changing magnetism. The declination is changing year by year and maps printed today become out of date quickly. The north magnetic pole is not just moving it is accelerating. At the beginning of the 20th century, the north magnetic pole was heading north-northwest across Canada at 10 kilometers per year. By about 2000 it was moving at 40 kilometers per year, and by 2016 it was up to 55 kilometers per year. The strength of the field likewise varies. Even if magnetism was found in the specimens when they were exposed in excavations, it might have been just another very confusing batch of data to muddy up the work. Stopping the effort at

Magie's admonition was probably a wise move by Moore which saved money and time.

The results of Fay's magnetic survey of Meteor Crater as of the beginning of October 1920 suggest four or five magnetic fields are at the crater. The strongest of the fields is on the southeast side. The men in Boston are told this, but it does not change anything. They are many weeks into site preparation on the arch of the south rim. And that site was well away from the suggested magnetic anomaly. But the southeast anomaly is considered as the possible location for the second drill site.

They had some magnetic indications but mostly the physical uplift of the arch to suggest it was the best first location. This had been a long-held view of Barringer and Magie, and it is hard not to believe that they strongly influenced the decision by the United States Smelting Refining and Mining executives. Barringer had nearly twenty years of time to work up his arguments and defenses that the asteroid's resting place was beneath the south cliffs.

Mr. Fay was nearly done with the survey by October 8th. This survey has not been as elaborate as what Barringer or Moore had proposed. Fay was unable to get sighting at many of the locations that Barringer's survey would have required. Holland stated that Barringer's survey would have consumed some months to complete. Barringer wrote in a letter to Mr. Moore on August 19th that the survey should take just three or four days at the most. He says further that depending on the results they can at a later date check the work with a delicate dip needle instrument. Barringer adds a suggestion that Professors Magie or Thomson can be of assistance in getting such an instrument. He further adds that the careful interpretation of the survey will be useful in locating the second drill site. Rather than taking 3-4 days the survey actually took seven weeks and was still not as detailed as everyone wanted. There is no reasonable explanation for Barringer's estimate of 3-4 days for such a survey it is simply an idiotic statement. His gross underestimation of the time may have been a tactic to make getting the survey done an easier sell to the men in Boston. Barringer was himself most eager to have another more detailed survey. He also wanted the survey to focus on the southern portion of the crater which was not what Moore wanted and not what he had instructed Holland and Fay to do.

Holland writes that he has followed the ideas clearly expressed by Mr. Moore and he has made sure they were followed by the surveyor. Mr. Moore was a vice president and consulting engineer back east, and Barringer was not Mr. Holland's boss at all. Though he was to extend courtesy to Barringer and accommodate him, Holland knew who he was working for. This report was written to Sidney Jennings another vice president and A. P. Anderson another consulting engineer. It is worded carefully to ensure they all know that Holland is clear who he is working for and whose money he is spending. But, Holland is told in a later correspondence with Moore once the magnetic map has been studied that he actually failed to follow Moore's instructions also. Moore does seem to realize that the survey he proposed would have required enormous time and expense. He sort of lets Holland off the hook accepting the survey grudgingly. But no one is happy with the finished survey. Mr. Fay had already been paid either nine hundred or one thousand dollars for this survey. Holland reports both figures but without exact dates, so both may be correct depending on when during the survey Holland is reporting about the expense. Holland will also take some flack from Barringer who criticizes the survey for not focusing finely enough on the south rim. Barringer had wanted a survey where the observations were taken at intervals of just 200 feet and when anomalies were found to trace them in all directions until the reading was again normal. Each anomaly was to be a red circle with arrows indicating magnetic deflection. Each normal reading was to be a black circle. The pattern of red and black circles was to form an easy visual indication of the location of an iron mass. This could have involved hundreds of additional stations with three readings during the day at each.

It is never discussed whether Mr. Fay moved the transit from location to location during the day or stayed at a single location and exact set up to take his three readings. If the latter is true, then it is easy to see why the survey would take months. Only one station would be recorded each day. This would seem to be impractical. Mr. Fay was considered by all involved to be a highly qualified surveyor. This would mean that he was able to place, level and measure with his transit repeatedly with accuracy. It would seem to be almost part of the definition of a surveyor that he be able to get repeatable results. However, there is no mention of whether with such small variations he may have chosen a hands-off approach and left the instrument in place unmoved for the three daily readings. It would eliminate one source of human error in the data. After seeing problems with moving the instrument and not getting repeatable results he may even have felt forced to adopt the method of leaving the instrument in place for the three readings. Moore produced a sketch of the crater and the surrounding area with points marked that he felt should be used as a pattern for the survey. Holland gave a copy of this sketch to Mr. Fay to follow in his work. Moore wanted readings taken along the cardinal compass directions at intervals out to a mile from the crater center. It was Mr. Moore who requested the survey include stations "not only all around the rim but to cover the inside or pit of the crater as well. These reading, of course, need not be taken as close as 200' centers. . . . If an appreciable variation should be found at any point, then readings around this point should be taken to cover the whole magnetic field." There is clearly now a problem for Mr. Fay and Holland. Barringer and Moore have discussed in letters forwarded to Holland that the stations will be on 200' centers. Now Moore describing his sketch says they do not all need to be 200' on centers. Also, the terrain of the crater does not allow for readings to be taken at every place falling two hundred feet from the last station. Outcrops with steep edges, knolls, valleys, gulleys, and boulders may all be in the way of the surveyor setting up his transit and getting sightings when following such a 200' spacing plan. Then there is the expense, Mr. Fay wrote in mid-December after the report was submitted and criticized that he would still have been at the work if he had done it to the extent Moore and Barringer had asked.

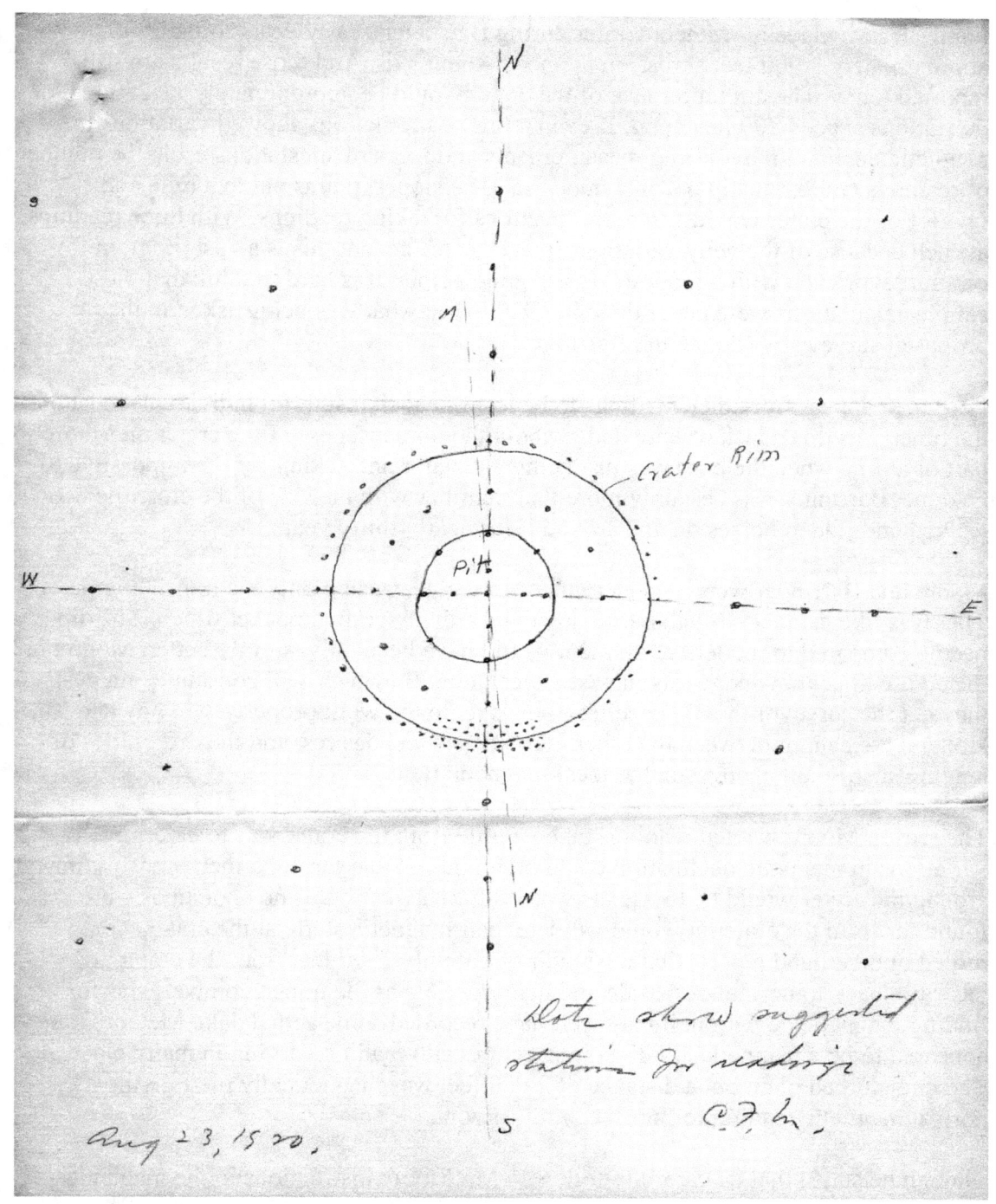

Pictured here is the Moore sketch of his proposal for the magnetic study. There are 120 locations where he wanted reading to be taken. It is clear that he has attempted to incorporate the wishes of D. M. Barringer for there are 40 locations near the south rim drill site area. That is the portion of the crater where Barringer was most interested in getting readings.

Land surveys made of Meteor Crater during Barringer's early years found its diameter approximately 3,900 feet across on the north-south axis. And those results are still reported today. The circumference of the crater would be approximately 12,250 feet or 60 stations spaced 200 feet apart. If even a few of the stations showed variations in magnetic declination requiring measurements at dozens of substations each, the number of readings could quickly become hundreds. The slopes, plains out to a mile and interior of the crater would add more locations for taking readings. With three readings at each because of the daily variations, it is easy to see that this is a vast program for one surveyor even with a reportedly very good helper. It is hard to think that these simple calculations were never done in 1920 to see what was being asked in the proposed surveys by Moore and Barringer.

Both men appear from their writings to be level-headed insightful individuals most of the time. Was this a lack of knowledge about the immenseness of the crater on Moore's part or a time when the boss was just being a dictator and asking for the impossible to be done. Barringer was certainly more than familiar with the size of the crater he had walked and ridden horses on and around it for nearly thirty years.

Moore and Barringer were also clear that Mr. Fay was using only the transit compass. That was like using a sledgehammer to create a finely crafted pocket watch. The dip needle compared to modern equipment would have been only slightly better. Moore, in the end, will acknowledge this flawed expectation. Barringer will constantly suggest they get the surveyor the dip needle he needs to do the work properly. Mr. Fay and Mr. Holland are caught between the other men's hopes and desires, and the difficulties of reading such small changes in the local magnetic field.

The reason Moore wanted readings taken a mile from the crater was to determine if variations in magnetic declination could be found out that far. If so then the work in and around the crater would be to use his words "for naught" but if no anomalies were found far from the crater, it would validate their magnetic study at the crater. As it turned out, Holland reports that the readings out only 2000 feet from the crater are basically back to normal. As crude and inappropriate as the transit compass was for finding magnetic deviation, it may well have recorded some actual data. Meteor Crater appeared to be a magnetically diverse structure with reading varying in many places. The undisturbed plain out a distance of 2,000 feet was magnetically inert giving a normal magnetic reading for north central Arizona.

Though he stated that he carefully followed Mr. Moore instructions in the magnetic survey Holland did not. Even as early as August 26, 1920, Mr. Moore is questioning how the readings are being taken and recorded. In both a telegram and a letter on that date Moore is seeking information about the "additional observations" that Holland had said are being made. It seems that they are not being marked on the survey map or are not readings being taken at the small distance intervals Moore wants. In the letter, Mr.

Moore describes the additional observations Barringer requested, and it is clear that he endorses this work. The stations appear on the map as either a varying declination or a zero difference from the magnetic north reading. The spacing between stations is very large. This is clearly not acceptable to either Mr. Moore or Barringer. They want the survey done so that when a variation in declination is found the instrument is moved slightly and more readings are taken until the reading fades and the declination is normal. These are instructions to perform detailed mapping of an anomaly to determine its size and shape. Moore states that he can not believe that a declination can be read at only the exact single spot and that it would not also show a reading if moved "say 10 feet from one of the stations where a declination was observed."

The actual lack of testing stations marked on the map does not seem to be deliberate shortcutting by Mr. Fay. Instead, it is that the requests by Barringer and Moore are beyond what it is possible for one man and a helper to do in a reasonable amount of time with the equipment they have.

The following is how Moore describes the survey he wishes to have done. "To be of any real service the field where declination readings are observed should be completely surveyed and marked on the map. The map would show where the maximum readings are and would show just how they fade out to zero and the directions which the magnetic lines of force take. It is very important that you get a hold of Fay and have this work done as quickly as possible."

Moore's expectations seem very reasonable. Very little would be learned from a map that only shows a single spot here and there with a declination other than to the magnetic north. A map with field strength lines at small distance intervals around any declination anomaly would form a series of contours encircling the places where disturbing iron might be. This is the kind of map that would be produced today in a survey, and it was a method understood even in 1920. The finished chart would resemble the contour map of hilly terrain only the contours would be magnetic declination variations, not elevation lines.

Moore finished this letter with a request that "a reading should be taken at the bottom of the shaft, regardless of the iron which is there. I understand there is a good ladder in the shaft and there will be no difficulty involved in making the reading." No difficulty other than getting yourself up and down a 155-foot deep hole with a heavy transit compass and tripod. Surely Mr. Fay had assistance in handling the instrument, and the reading was taken. It was also the strongest reading made. It is questionable if it was a reliable reading, though. There was a good amount of iron in the shaft that could easily affect the compass. If the hole had been cleared of iron, the reading might have been trustworthy and quite important since it was taken at depth. The iron meteorite fragments struck by the drills on the floor appear to be numerous at just a couple hundred feet deeper than the central shaft. Again a dip needle device would have been of value more so then the transit compass. In a post script to the letter written on August

27th Barringer admits to not having known about iron pipe being in the bottom of the central shaft. He is aware that it would affect the readings he suggested be taken there. Barringer is clearly the instigator of the request for a reading in the central shaft. He had used a simple dip needle down there years before and it seemed to indicate a magnetic attraction to the south. He confesses that the reading "was so slight and the instrument was so lacking in delicacy that I did not care to go on record as stating it to be a fact. This has been the reason why I wished Mr. Fay to take an observation in the bottom of the shaft." Twice in this letter, Barringer makes mention of his strong desire to obtain by some means a delicate dip needle instrument for Mr. Fay.

The survey map was finally prepared and sent off to a long list of recipients in October of 1920. Professors Magie and Thomson are on the list. By mid-December Holland has heard from his bosses in Boston that the map has never arrived. Holland sent the charts and maps directly from the printer to all the men. Holland writes a letter to Elihu Thomson on December 16, 1920, asking if he received the Magnetic Survey Map. He tells Thomson that the copies which were supposed to be sent to the Boston office have gone astray. A letter written by C. F. Moore on December 13 is already on its way to the crater stating that the map was found. It had been mislaid in the Boston offices. Holland is instructed to go ahead and have the additional copies made and sent as they will have use of an extra in Boston. This is the letter which is critical of Mr. Fay's survey. Since they just found the map they have been able to study it finally. They are immediately dissatisfied. It was not as systematic as requested and did not have readings at substations around locations with variations in the magnetic declination. Mr. Moore concludes "the results as given by the map amount to little. However, not a great deal was expected as a transit compass needle is obviously not an accurate instrument for work of this kind." The survey took seven weeks and cost a great deal of money and no one attempted to provide Mr. Fay with the proper instrument for the work that they knew he needed and which Barringer all but begged them to get him. In the end, none of the men in Boston felt there was much that could be learned from the survey. Neither of the notable professors Magie or Thomson thought there was anything definitive that could be learned from the data. Magie was intrigued by the anomaly Holland calls "freakish" to the west of Station 8. These two professors would be Barringer's most faithful supporters as the controversy about Meteor Crater reaches its most dramatic phase a few years later with the final shaft on the crater's south flank.

Holland was the supervisor of all the work done at the crater for United States Smelting Refining and Mining Co. While the magnetic survey was being done by Mr. Fay and his helper they were under his watch. Holland seems to have tried to do what he could to get as much data as possible in a reasonable amount of time. Holland recognized far better than the men in Boston or even Barringer that the requests they made were beyond the scope of a quick study with the wrong equipment. He was also up to his neck with other work. The pipeline was being installed with great difficulty; the drill site was being blasted and cleared. He had continuous labor problems; the camp was not ready for men to live in during part of the time. It seems likely that he had to leave Mr. Fay to his work and check on his progress occasionally.

It is possible that no real data was recorded on the survey map or in the surveyor's reports and notes. But there remain the same reasons seen by the individuals in 1920 to believe that real data on the magnetic characteristics of Meteor Crater was discovered. Perhaps the best support of the correctness of the survey is that the readings out on the surrounding plains were normal magnetic declinations. The fact that the readings of the crater floor were erratic and that much-abandoned metal was there makes a statement that the compass was, in fact, sensitive enough to "see" the attraction of local metal. This also would seem to make a statement in favor of the care and skill which Mr. Fay took in making the readings. Later studies with electronic equipment in the late 1920's and 1930's would give results that were all over the place and never agreed about where anomalies were. Studies done for H. H. Nininger in the 1940's would not give definitive results either. Aerial magnetometer studies even more recently showed little evidence at the crater and actually showed an anomaly unrelated to Meteor Crater some miles away. The study by Mr. Fay was as good as any given the circumstances and the tools. It was better than the study done by Gilbert with a dip needle thirty years before.

What should have been the expectations for the study? They knew that the buried asteroid was more than 1000 feet below the rim if it was there at all. The meteoritic material discovered below the crater floor was hit between 400-680 feet by the drills. If you add the height of the crater wall of 500 feet, then you see where the drillers under Holland's supervision expected to hit the asteroid. Barringer in letters about the magnetic survey expected the meteorite material to be highly oxidized at least on the outside. What kind of magnetic attraction would highly oxidized nickel-iron have compared to solid metal. Much less certainly and would be harder to observe with the compass needle. When the drill finally got below the 1,300-foot level, the bailer began to show undisputed meteoritic material which Barringer is often quoted in saying was "the rottenest material" he had yet seen. Could the compass more than a thousand feet away have recorded any more than the tiny amount Mr. Fay reported? Maybe the data was truly the best that was obtainable for the time and instrument used.

Holland's final remark about the survey is in response to Elihu Thomson's discouraging analysis of the data. Holland writes "they will have to rely on the drill to find the asteroid." In just a few days after the survey maps are mailed, he will begin drilling. By the time Boston finally sees the magnetic survey map in December, he is well into the troubles up on top of the south rim. Mr. Holland's real nightmare has begun.

The Drilling

It has not been much of a Christmas Day for me so far. Maybe after I write my weekly report, I can take some time to relax. I could write a note to Lillian, she has been home for a few days, and I have not written her since she left the crater. Some of the men have gone to California for the holidays, and we are covering for them with 12-hour shifts on the rig. I could use a break and some time away from here, but I have too much going on and too many problems. I cannot believe this horrible equipment; it is like a curse. Another drill stem broke this morning and must go to Los Angeles for repair. One broke last week at the same place on the rod. The steel is no good, and they are too small a diameter. Even so, we should not be breaking four-inch steel drill rods on a daily basis. I fear that even when repaired, they will be no good and will break again. Now we are trying to drill with a tool string that is too light. I guess I may have to wait until next year for a peaceful and relaxed holiday season. I have to say though the weather has been spectacular, nights with frost and daytimes that are sparklingly clear and bright. I wish Lillian could have stayed to see them. If it were not for the lost tools in the hole and the breaking stems I could wish to stay here for the beauty of the area and the mystery of the crater. I hope we can get past these problems and finish this job. I really want to find a buried star.

Despite all the other things Holland did at Meteor Crater in the months from May to October 1920 it was a drilling program. That drilling finally began on November 1, 1920, when they "spudded in" using the driller's jargon. They ran the rig three shifts a day of eight hours each. They called the work shifts "towers" more jargon. The men began with high hopes. But, some individuals expected it was going to be difficult. Barringer already on November fourth wrote about the crevices and cracks possibly being a problem. In the same letter, he mentions the hard obstruction struck by the drill at the corner of shaft number three on the crater floor. He writes Holland that they blasted the obstruction to no avail and when they moved the hole just three feet over they did not again hit the blockage. Holland would never during his time on the job hit any of those type obstructions, but he would have plenty of problems with machinery, cracks, and crevices.

A summary of Holland's weekly reports seems the best way to get a feel for the work and the hardships the drillers experienced.

On October 31, 1920, Holland made his last report before drilling began the next morning. They had nothing but troubles during the week as seen by what he wrote. "In my last weekly report, I stated that the plant was just about ready to run if no unforeseen difficulties arose. Defects in the gasoline engine and its accessories have given a great deal of trouble, with the result that though the drilling crew have worked like draft horses, there is very little to show for it. Our location is, of course, very unhandy for emergency supplies of any kind. In view of the possibility of difficulties in starting up the gasoline engine, I engaged the services of a gas engine expert who is well recommended in Flagstaff. He did not arrive until Wednesday and could do little to help during the time he was here. However, the engine is running much better today and starts with less difficulty. Tomorrow (Monday) it will be safe to put on the three regular towers for steady drilling."

The drilling does commence on November first. Holland's reports are done weekly, so his next report is on November 6th. They make a good beginning but even this early he has problems to report. "Drilling on three towers started on November 1st. By November 3rd. The hole was down 97 feet from the derrick floor. The formation penetrated is entirely limestone so full of crevices that the tool continually has a tendency to drill out of line, and it has taken the rest of the week to get the hole in shape to take the ten-inch casing. There will be approximately 250 feet of the limestone to go through. It may be less fissured with increasing depth, but the hole will no doubt require special care that it does not get crooked in the limestone. The underlying sandstone should be much easier."

One of the few times he reports on equipment issues other than the drill tools themselves is in the next paragraph. "The blower sent with the drilling equipment is intended for operation with steam and the arrangement substituted for running the blower from the drill engine was unsatisfactory and dangerous. A small gasoline engine borrowed from Mr. Hart is now running the blower very well. While we have so far refrained from "taking the shirt off his back" we have at all times taken anything of Mr. Hart's that we needed, at his cheerful invitation."

There is no mention of any arrangement that Barringer had with the local cattleman Mr. Hart. He is mentioned regularly and always as a very helpful and interested person in the work at the crater. It is strongly suspected by this author that there was to be some form of compensation at a later date if the asteroid was found. It seems almost certain that Barringer had some agreement with Mr. Hart.

By the next week, the early fast progress has stopped, and the new pattern of fighting the broken rocks has already begun. In his report of November 13, Holland discusses no other topic than the drilling. The pipeline is fully complete and is no longer an issue. The post office to be located at Sunshine is moving through the federal government bureaucracy and not brought up this week. After two weeks of drilling, this report is the first of many to follow that will sound similar.

"The entire work has been occupied in straightening crooks in the drill hole, caused by the bit persistently following crevices in the limestone. Tonight the hole is straight, and the ten-inch casing is free to the bottom, but it has required all the driller's ingenuity and patience to make it so. Fortunately, the drillers, especially the foreman, have had considerable experience in drilling fissured limestone elsewhere. They do not anticipate the same difficulty in drilling the underlying sandstone, even if it is cracked and fissured. As may be expected on the axis of the uplifted dome, as the sandstone will be much less hard than the limestone, and easier to get a "shoulder" on if crevices are encountered. The crevices in the limestone are of such size that a lighted lantern lowered into the hole will go completely out of sight."

Holland again reports briefly on the gasoline engine and the difficulties they continue to have with it.

"The "Ideal" gasoline engine is very hard to start, especially when cold and its consumption of gasoline is very high. I have taken up these matters with the manufacturers, in the hope of getting some relief."

The quote marks that Holland places around the word Ideal seem to be a bit of sarcastic humor on his part. He never wanted the gasoline engine and had several discussions with Boston about the engine. His experience was with steam. It might not be beyond the truth to see a poke at his bosses in his quote marks around the word Ideal which was the manufacturer's name.

With some problems taken care of and the drillers aware of the behavior of the tools in the fissured rock they have made some progress by the November 20 report. Holland's wording is sometimes hard to read nearly a hundred years later. He will often use the word "instant" to describe what is a current status or to mark a moment of time when he receives a message or says something in a message. "On Thursday 18th instant the drill hole entered the white sandstone at 195 feet from the derrick floor. The limestone was therefore somewhat thinner than Mr. Barringer's estimate. The sandstone was penetrated for a few feet, but as the hole did not hold water at all and it was caving somewhat, it was decided to put in the eight-inch casing and substitute an eight inch bit for the ten inch. Yesterday (19th) the gasoline engine was hung up on account of the breakage of two essential bearing studs which could not be replaced locally, but they are expected to arrive from Los Angeles tomorrow, Sunday. In the meantime the crew are working on the tools, making improvements in the rig and appliances, and putting in safety rails around the main belt."

They have made it down to 195 feet in less than three weeks of drilling. They are not going fast, but they are making progress. He seems hopeful at this moment that the engine will be repaired promptly so drilling may commence again. He knows that if he does not tell Boston what the men are doing during this downtime, they will ask, so he gives the list of tasks keeping them busy. The tool dressing would be going on regularly as the bit got dull and wore down. The mention of the safety rail is interesting. Today we often consider the industrial worker of 1920 as someone working in poor and hazardous conditions. And this may have often been the case. Yet repeatedly at the crater, concern for the worker's safety is shown by both Holland and Barringer.

The hole not holding water was probably the most important of the concerns in this week's report besides the engine breakdown. In churn drilling, the hole needs to keep water in the bottom so that the pulverized rock powder can be removed by lowering a bailer which will fill with the mud the bit creates. Without water the bit pounds on a layer of rock powder and not on the rock itself and little or no progress is possible. The solution to a dry hole was to dump material down from the surface to plug the bottom so it would hold water. Clay, rock, even cement might be used by drillers to prevent the water running out the bottom.

November 27 Holland writes what was surely a difficult weekly report. "I regret to report that no progress has been made in deepening the hole during the past week. The parts to replace the breakages in the gasoline engine, ordered in Los Angeles by wire on the 18th. instant and said to have been shipped by express on the same day, did not arrive at Canyon Diablo until Wednesday the 24th, in spite of many telegrams and tracers. The engine was running again in less than an hour after receipt of the parts at the Crater, but it was then found that during the shut down of the power, the casing had become tightly cemented in the drill hole, probably owing to the packing and cementing of caved lime. Various methods, including the blasting with torpedoes, have been tried to loosen the casing, without success so far. The drillers are now preparing to shoot off the casing shoe, and perhaps the bottom section of casing, in the determination to get it free, and to resume drilling."

We receive some interesting insights into 1920 in this report. There were no transcontinental flights or computer controlled shipping centers. However, there was express shipping and same day delivery of packages in that time. There were means of tracing and tracking packages as there is today. And there was clearly just as much frustration as today on the part of the receiver when a package failed to arrive on time. But there was nothing Holland could do out at the crater in the middle of Arizona to get the engine parts there faster. He was at the mercy of the company in Los Angeles and could only take their word that they had shipped when they said. In the end, the delay was a major setback for the drilling. It created problems that were never fully resolved during Holland's time on the job.

On December 4th Holland writes two reports and does a written inventory. His direct boss Sidney Jennings was at Meteor Crater for a visit the day before. The drill was still not running and the casing was still cemented in the hole the day of his visit. Here are portions of those reports. To Sidney Jennings, he writes "Since your visit yesterday the hole has been redrilled to a depth of 160 feet. While the progress has been slow it is as fast as our foreman thinks safe under present conditions. There has been a reduction in the consumption of gasoline. The engine has run continuously, but during the past twenty-four hours it has consumed only 80 gallons. This is much nearer the makers' estimate of 72 gallons per day than the three barrels it has been consuming until the various adjustments in the engine were made."

To Mr. A. P. Anderson his other boss he writes a complete report of the whole week.

"Immediately after my last weekly report was made the shoe was shot off the casing, and later about fourteen feet of the bottom joint, in order to get the casing free. After it became cemented in the lime during the stoppage for repairs of the gasoline engine. The remaining casing- fourteen feet- has been drilled out and the hole is redrilled to its depth of 160 feet, so that 40 feet more have to be redrilled to reach the bottom of the hole. Progress is rather slow, but probably as fast as should be made for safety under present conditions."

He continues with additional information about the repairs they have made to the gasoline engine mentioning an untrue shaft that may have been responsible for the excessive gasoline consumption. What we learn from this report and not from the report to Jennings is that the casing shoe had been shot off days before Jenning's visit and that only the 14 feet were blasted off right after he left the crater. The drillers have broken up the casing with the bit and brought up much of the metal fragments in the bailer. But they will continue to have metal in the hole and coming up in the bailer for a long time. This remaining metal may be the reason for the slow progress in redrilling the hole and the concern for safety.

As careful as they were being they have more problems in just two days which further slow the redrilling. Holland shares these new difficulties with Boston in his report on December 11.

"On Monday night 6th instant, the drilling jars broke and the tools were temporarily lost in the hole. The fishing tools supplied by the Union Tool company through the Republic Supply Company, are not suitable for use in such an emergency, so it was necessary to fashion a suitable tool on the ground, by means of which the drilling tools were recovered."

"In anticipation of such a breakage, an extra set of drilling jars were ordered from the Republic Supply Company on November 27. They are said to have been shipped to Sunshine on December 1. Immediately after the accident, still another set was ordered by wire to be sent by Express, and these are said to have been expressed to Canyon Diablo on 7th December. All trains have been met but so far neither set has arrived."

"In the meantime, the reaming of the crooked hole is being continued with the long stroke jars belonging to the fishing tools. Though it is risky to use them for such a purpose, on account of their weakness, this is the only alternative to remaining idle until the proper drilling jars arrive. Tonight the hole has been re-drilled to a depth of 174 feet."

Jars is the strange name for a short part used in the string of tools which could be as much as fifty feet long. Holland preferred a tool string that was 30 feet long when he had enough parts to make one up. The "jars" were not containers as we would commonly use the word. The name refers to a different meaning of the word jars; to move something through striking it. The early drilling jars consisted of two interconnected links of steel that could slide within each other. After the drill bit fell upon the rock and crushed it, the bit would be pulled up again by the action of the

walking beam. The jars would allow the cable to rise freely for a few inches. The amount of free motion was "the stroke" of the jars. The links would finally lock, and the bit would be jerked upward. The jars prevented the bit from getting stuck in the rock it was pounding. But it is easy to see that "jars" would take a tremendous amount of stress. The total weight of the drill string was well over a ton, and the jars pulled that full weight with a jerking action thousands of times an hour. Later jar designs were more concentric and tubular devices but they served the same purpose and considering the number of different fishing tools created to retrieve them all designs broke regularly.

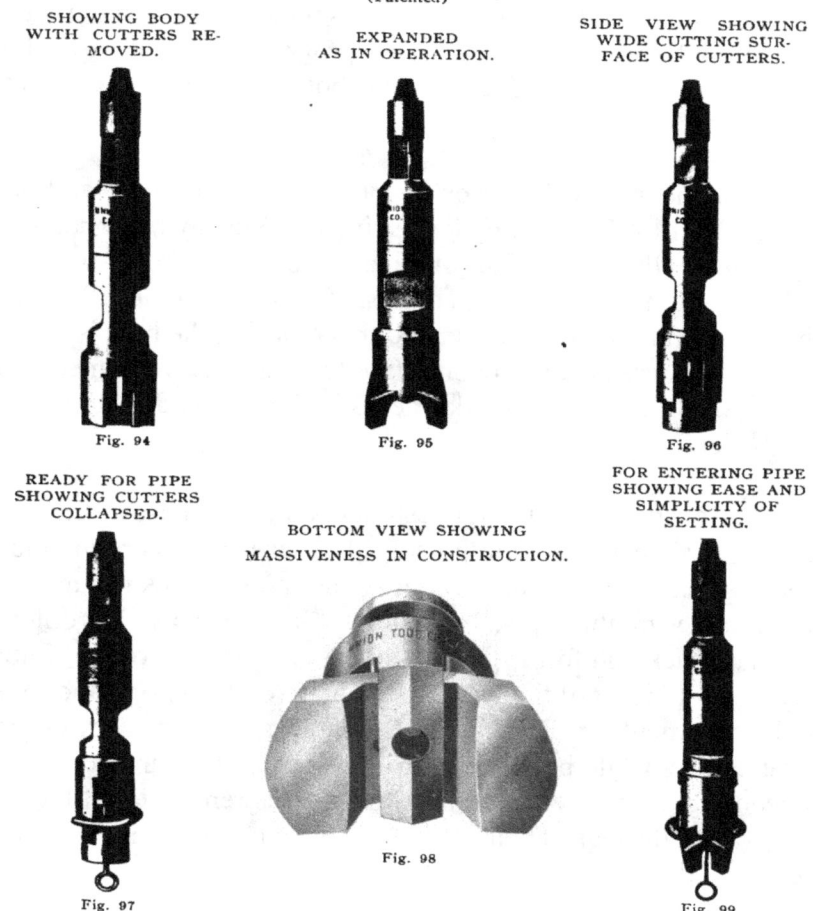

Drilling Tools
FOR CABLE SYSTEM

DOUBLE IMPROVED UNDER REAMERS
(Patented)

SHOWING BODY WITH CUTTERS REMOVED.
Fig. 94

EXPANDED AS IN OPERATION.
Fig. 95

SIDE VIEW SHOWING WIDE CUTTING SURFACE OF CUTTERS.
Fig. 96

READY FOR PIPE SHOWING CUTTERS COLLAPSED.
Fig. 97

BOTTOM VIEW SHOWING MASSIVENESS IN CONSTRUCTION.
Fig. 98

FOR ENTERING PIPE SHOWING EASE AND SIMPLICITY OF SETTING.
Fig. 99

FOR DESCRIPTION, PRICES AND WEIGHTS, SEE PAGE 43.

There is an important word used in the last paragraph of this report. Holland writes that the "reaming" of the crooked hole is continuing. The drill hole had to be made straight and also be the correct size for the casing. Often this meant using a bit that enlarged the hole diameter. It was called a "underreamer" bit. Underreamer bits came in several types. The illustration above from the 1914 Fairbanks Morse Catalogue shows one type of spring-loaded underreamer bit. The detail drawing in the center has Union Tool Company stamped on it. Union Tool was the Torrance, California based company with which Holland had so many difficulties. Spring-loaded under reamer bits were adjusted to expand to a specific diameter after clearing the end of the casing. They would close down when pulled back into the casing. Other underreamers were a fixed diameter, and the casing had to be withdrawn from the hole for their use. The type and style of underreamer used is not known. However, in a listing for an underreamer named the "Wilson" it is mentioned as a selling point that it is a single piece forging of nickel steel that will not break in two. In the illustration above the underreamer shows the thinner metal in the middle and the metal cut aways on both sides. Was it a underreamer similar to this shown that broke in half at Meteor Crater? Underreamers like the cutting bits were very heavy. A 10-inch underreamer is listed as weighing 690 pounds. The 8 ½ inch version was 500 pounds. More will be heard of underreamers in the future reports made by Holland. They have made just 14 feet of progress in redrilling the hole this week, and they are still some 20 odd feet from the bottom that was reached earlier by the bit.

Holland's next correspondence with Boston is not on his regular interval. It is only three days later on December 14. "The drilling jars arrived on Sunday by express. They are of the heavier type generally used in California and should not break so easily as those which came with the original equipment. The casing went the whole 200 feet to the bottom of the hole last night in good shape, and this morning the hole is 218 feet deep. The white sandstone bed was apparently only five feet thick as we immediately got in limestone again, beyond the 200 foot mark. Pieces of the casing drilled out are still coming up in the bailer."

The earlier idea that Barringer had wrongly estimated the thickness of the Kaibab Limestone is proven incorrect. The limestone can be measured all around the crater, and its thickness was well known. However, none of the main rock groups is monolithic as their names might imply. The Kaibab Limestone also has dolomites layers which alternate back and forth. The massive alpha member of the Kaibab is certainly different from the rest of the Kaibab below it. More of hope and relief appears to have been in the statement that Barringer's estimate was wrong. They desperately prayed that the sandstone would be easier to drill. Reaching the sandstone was a major goal. They had been tricked by the crater and did have the remainder of the known 250 feet of limestone to fight through. Their 18 feet in a single night made reaching the end of the limestone a goal they thought could be soon achieved.

Holland makes a report on December 18th, back on his normal schedule. Some of it is a repeat of what he wrote in his midweek report. "A set of jars arrived by express on Sunday and immediately replaced the long stroke fishing jars then in use."

"At 200 feet the bit struck limestone again, so that the sandstone bed was only five feet thick, from 195' to 200'. Drilling continued in limestone to 250 feet, when the stem, 30 feet long and 4 inches in diameter, broke immediately above the box. The under-reamer being used at the time was lost at the clean horizontal break, and the crew have been continuously attempting to recover it since. As in the case of the broken jars, the fishing tools have not been of much use for the purpose. The impression block showed only two small dents on its circumference, indicating that the tools have fallen sideways out of line with the hole, probably in some soft material."

Holland inserts a simple drawing of the impression block at this point. Then he continues with his report. "We are now "drilling past" in hope of getting the casing over the tools or otherwise straightening them for recovery. The broken stem has been taken to Sunshine for shipment to California for repair. In the meantime the shorter stem must be used."

They were making some good headway on the hole until they are again hit with difficulties. They are down 250 feet and getting near the end of the limestone a point maybe only 20 feet away. But the long drill stem to which the bit and jars are screwed broke. It was 4 inches in diameter and weighed hundreds of pounds. It is hard to think that steel that thick would break clean off. There will soon be a pattern of similar breaks indicating that the drill stems are defective either in design or manufacture or both. They have been called wartime products in writings by Barringer and others. They may have been manufactured when most good steels were devoted to the war, and lower grade steel was reserved for domestic efforts. There was a tremendous upsurge in steel production during the years of the First World War. It had returned to prewar levels by 1920. It is possible that the drill rod sent to the crater was made of surplus wartime material.

The underreamer is not blocking the hole, but neither is it out of the way completely or contained behind the casing. They continue drilling down past it. Holland makes a logical conclusion about the lost bit which may well be true, believing it has fallen to the side of the hole into soft material. But because it is never recovered it will prove a disaster later. As long as lost equipment was held behind casing, it was considered safely contained. No one could have foreseen how fickle the rocks of Meteor Crater would prove to be. They almost regularly hit pockets of sand that would run into the hole. The running sand would bring along whatever was mixed with it; cobbles, boulders, or lost tools that have "fallen sideways out of line with the hole, probably in some soft material."

Holland's December 25th report certainly was one of the most difficult thus far for him to write. Worse reports were still ahead. The drilling program begins its move toward becoming a real disaster near this time. "This (Christmas) morning another stem broke above the box. At the same place as last week's broken stem, but at the top end. By 10 a.m. the stem and bit were fished out of the hole and drilling resumed. This stem will also have to go to California for repair. We still have an 18' sinker to drill with, but it is of course lighter than the regulation 30' stem."

"The week has been occupied with continuous efforts to recover the under-reamer, lost last week, or to side track it, without success so far. The hole is still in limestone at 257 feet, or 7 feet beyond the point where the stem broke. Dynamite has been used in the effort to move the under-reamer so that the casing will go down. The impression block shows that about half the area of the hole is obstructed. The bits become badly battered or broken in a short time, and there are many flakes of iron in the bailings. At 256' a few small pieces of shale ball came up in the bailer, but I found none today. One pair of drillers has gone to California for the holidays. The other two pairs are working twelve hour towers."

Holland's Christmas report offers many insights into the condition of the hole and to the extreme measures he and the men used to continue drilling. The hole was still partially blocked by the underreamer. The impression block he has referred to before and again here is a tool that could be screwed onto the bottom end of the drill string. It had a soft lead block that dropped onto the bottom of the hole or an obstruction. A dent or "impression" was made in the face of the lead block from which the drillers got an idea of how the hole was obstructed. The impression block now shows the hole about half blocked. The previous report had included Holland's drawing indicating just two small dents on the very edge of the impression block. Have their attempts to lower the casing caught the underreamer and pulled it further into the hole or has it slipped into the hole on its own? It is impossible for us to know but the underreamer is now being hit by the bit and bailer as they move up and down the hole. Additionally, the cutting bit is constantly being damaged by pounding on chunks of steel.

They are just getting to the level where meteoritic material begins to appear in the contents of the bailer. But the small amount of shaleball meteorite is soon passed. This was likely a pocket of mixed debris filling a crevice containing some nodules of shaleball meteorite.

Two stems have broken in a week, and even though they were different lengths, they have broken at the same location on the stem. The box is a forged shape on the ends of the rods. It is easy to see that it could be a place of weakness and it has proven to be so. The drillers have to use a shorter stem only 18' long which is much lighter in weight. The pounding through the rock will be slower.

There has never been a clear statement about the number of men that were recommended and hired on the drilling crew. We know that Mr. Coan of the Oak Ridge Oil Company recommended Mr. Wammock as the foreman. We also know that Mr.

Wammock selected his crew and that Holland had put the men on the payroll. But until this report, we are not told how many men came from California. There were two men per shift, and we know that Mr. Wammock had agreed to work a shift instead of being a supervisor only. It is reasonable to conclude that the crew was at least five additional men besides Mr. Wammock. Small vehicle mounted churn drills still used today can be run by just a single man, but two men were working each shift at Meteor Crater. We know though it is not mentioned here that there was also at least Mr. Wommack's son as a tool dresser. It is likely there were more than one of these workmen as well. Perhaps one tool dresser per shift. The tool dresser was really keeping busy at this point repairing the battered bits.

Mr. Holland received a letter from Sidney Jennings on December 30 which was dated the 24th. This correspondence included an extract from a letter written by D. M. Barringer. Holland responds to this correspondence with a letter of reply the same day December 30. "I have the favor of December 24th. enclosing an extract from the letter of Mr. Barringer in which he refers to the water level in the Crater. I shall not fail to note, and keep you informed, of the action of the water in the drill hole."

"You will recall that the five foot bed of sandstone penetrated from 195 to 200 feet, drained the hole dry. From 200' to 265', which I regret to say is our present depth only, the formation has been limestone. At times somewhat siliceous, which retained water long enough to fill the bailer for ten feet or more, though the hole will run dry if water is not fed frequently. There are undoubtedly still crevices or caverns in the limestone which allow the water to run away below the casing."

"These crevices may be a contributing cause of the continued fatalities to the drill stem, though I am inclined to believe that the trouble is mainly a matter of poor war-time steel. I will give details of these breakages in my weekly report. Suffice to say for the present that all four of our drill stems and sinkers have broken in the same place, and that at the present moment there are in the hole an under-reamer upside down, and two bits."

The hole is down to 265 feet in depth, and they are breaking drill stems as fast as they can put them on. Holland makes the first mention of "poor war-time steel" at this point, and it will be repeated and republished for the next nine decades as something Barringer said. But, it originates here with Holland. The fact that all the stems have broken at the same location seems to indicate a bad design. The fact that four-inch diameter steel stems are breaking in limestone which while hard is not by any means as hard as other rocks would indicate poor manufacturing. They will send all these stems off to California to the Union Tool Company of Torrance for repair. The repairs will take so long that Holland will go to Los Angeles and find another forge to make stems and do repairs. He will choose the Regan Forge in San Pedro. How much actual work the Regan Forge provided to Crater Mining Company is unknown for the future will bring Holland and the drill program many changes.

Mr. Holland is not getting much time off for the holidays himself even if some of the drillers are. He wrote a report on Christmas, and he wrote a letter to Jennings on New Year's Eve about the sampling of the bailer contents and the storage of the samples. Now he is writing his weekly report on New Year's day. He will go into all the details of the previous week because though Jennings has already heard Mr. Anderson has not.

"Drilling. On 27th December a third sinker (16') broke in exactly the same manner as the first 30' stem, losing the bit in the hole. All three have been shipped to California for repair, leaving nothing to drill with but a ten foot sinker. This broke on the 30th. December in the same manner as the others, at the square next to the box. The last stem and bit were recovered the same night, but there are still in the hole the under-reamer and an 8 ¼" bit. The impression block indicates that the latter is lying horizontally on its side. A six inch bit goes down alongside of it at a depth of 265 feet, but of course the string of tools is very light to drill with, as there is no stem. Fishing for the under-reamer and bit will be continued as opportunity permits and additional fishing tools have been ordered to make up deficiencies in the present equipment. The Union Tool Company is investigating the breakages and has asked that the pins and boxes which broke off be shipped back. The only plausible explanation, however, would appear to be that the stems are a very inferior war-time product. Although no consolation is to be obtained from the knowledge, it is a fact that the local Division of the Santa Fe Railroad is having an extraordinary number of breaks on the shafts of their new locomotives, perhaps for the same reason."

"The last impression obtained of the under-reamer indicated that it is upside down. Nothing of it shows on the impression block now, There is evidently a large crevice or cavern in the limestone here. According to Fay's profile map of the MN line it would need a tunnel of prohibitive distance of 280 feet from inside the Crater to reach the present bottom of the drill hole."

Things are dreadful now. Every stem or sinker Holland has to work with is broken. He can put the bit almost directly on the end of the cable but it is too light to drill with and only a six inch bit will go all the way to the bottom. There is a lot of steel lost in the hole. The underreamer and the 8¼" bit are lost. The bit was not new and so was shorter but it undoubtedly still weighed many hundreds of pounds. The underreamer was upside down. That made it very hard if not impossible to pull up through the casing. Even though it did not show as an obstruction on the last impression, it was still down the hole.

The first suggestion of a tunnel to recover the lost tools is made here by Holland. It is clear that he does not like the idea and considers it too long a distance to dig. He does not realize yet that the tunnel will soon be the focus of everyone's thoughts back on the east coast. It will also end up being far longer than the 280-foot estimate written here.

The weekly report of January 8, 1921, is a mixture of bad new and small progress. "The week has been spent in alternately "fishing" and "drilling by", but with no stem on hand and inadequate fishing tools, little practical progress has been made. The hole is now at 172' deep. So it has progressed only a foot a day from the last week. It is still in limestone. There is considerable iron from the tools in the bailings."

"The first broken stem was shipped from Sunshine to the Union Tool Company on December 21st. The Republic Supply Company, through whom all our equipment was obtained, wired that the stem was shipped back on December 30th but it did not actually arrive until tonight (Saturday) and then was sent Freight Collect instead of Prepaid. Ordinarily it is customary to take unpaid Sunshine freight to Winslow until released, which would have delayed delivery of the stem at Sunshine for at least three days more, release has been obtained through the help of the agent at Canyon Diablo and the stem will be at the Crater on Sunday morning...."

Holland goes on to discuss the change of ownership of Union Tool Company and complain about the bad service and hopes it will improve with the new owner Nation Supply Company. There is one thing to note in this report, and that is the depth that he reports for the hole. It is incorrect and will be corrected by him later. It should be 272', not 172'.

It is at this point in the drilling that the rapid flurry of telegrams and letters occurs. Holland is receiving and replying to a message from Boston on almost a daily basis. A quick summary of these is as follows.

January 10th telegram from Sidney Jennings to Mr. Holland
Canyon Diablo, Arizona
"Referring your weekly report January first would it not be best to move the drill some distance to the east and get rid of the necessity of fishing out the tools that have broken in the present hole. Please let me know relative estimate of cost moving derrick sinking new hole two hundred sixty-five feet and fishing out tools."

Holland's reply on January 11 to Sidney Jennings from Canyon Diablo
55 Congress Street, Boston, Mass.
"Believe necessity of moving rig will be avoided by drilling past at less expense than moving which will be heavy. Will make estimates desired and wire them to you."

Day Letter Holland to Sidney Jennings January 13 from Winslow, Arizona
55 Congress Street, Boston, Mass.
"Holding back estimate of cost moving drilling outfit some distance east and redrilling until rig builder telegraphs his figures. In event of having to abandon present hole believe it will be comparatively easy and inexpensive to jack derrick back five feet, leaving the engine alone. None of fishing tools ordered yet arrived and progress is very unsatisfactory, especially as another bit was lost this week besides casing shoe."

To L.F.S. Holland Canyon Diablo, Arizona
January 14th New York, NY
"Your blue fourteen in event necessity to abandon present hole would prefer to have derrick moved five feet north or nearer to Crater than back. Will await with interest various estimates of cost."
Sidney J. Jennings

Day letter from Holland Canyon Diablo, Arizona January 14
Sidney Jennings 120 Broadway, New York
"Derrick so near edge that if moved five feet north main mud sill carrying walking beam will be off rock formation. But feasible with timbering at a cost of about nine hundred dollars complete by utilizing drilling crew and one rigger with jacking equipment Stop Cheaper to move south but hole might be less valuable Stop Past performances in limestone indicate new hole may cost seven dollars foot present depth two eighty two Stop Present tools probably adequate in underlying sandstone but to guard against repetition disasters safer to provide heavier stem and joints for limestone starting with twelve and half inch bit total cost tools approximately eight hundred Stop Riggers labor present derrick cost twenty seven hundred Stop Should wreck and rebuild for one half but may spoil considerable dry lumber Stop Hauling aggregates for engine foundation cost two hundred Stop Complete new material for another rig on ground Stop Two days after arrival overdue appropriate fishing tools about sufficient determine if hole can be saved."

Jennings to Holland from Boston January 17
Canyon Diablo, Arizona
"Your blue seventeenth if after fishing program decide that hole cannot be saved prefer to have derrick moved five feet north provided mud sill carrying walking beam can be made sufficiently solid. Approve suggestion use heavier stems and joints starting with twelve and one half inch bit."

Holland to Jennings, Boston, Mass January 22
"Two bits recovered rest side tracked. Drilling slowly in limestone having no stem."

During this exchange of telegrams, Mr. Holland makes his weekly report as well. On January 15 he sends a lengthy and tragic report of many small disasters and few successes. "A further series of breakages of drilling tools during the week and the nonarrival of any of the fishing tools urgently ordered, has resulted in small progress. Today the bit went to 282' and the bailer to 280'. (I have just observed that a typographical error in my last report showed the depth of the hole then 172'. It was, of course 272') A bit hook is said to have been expressed from San Pedro, California, yesterday and a horn socket (second hand) from Los Angeles today. A spud and other fishing tools ordered on December 29th are still to come."

"Immediately on its arrival at the Crater on Sunday 9th the 30 foot stem was hitched on.

On Monday 10th a 8 ¼" bit was lost from the jars and has not yet been recovered in spite of drilling by, and fishing with the inadequate tools available. On January 12th 8 ¼" casing shoe broke off. The casing had not gone beyond 262' 9", being obstructed by projecting lost tools, but was quite free and pulled out easily. A larger impression block was sent down after the casing was pulled. It showed little but the casing shoe, which also has not been recovered. On the night of the 14th the whole string of tools got stuck but was released by sending down a 6" bit and jars on the sand line. This bit went alongside and as far as the 8 ¼" drilling bit, so the hole or crevice is evidently large there. The lost tools appear to flop from side to side, allowing the drilling bit to go by but always projecting far enough to stop the casing from passing. As previously reported, the impression blocks have showed the lost under-reamer to be upside down, and a bit to be horizontal."

"This afternoon (15th) the fishing jars broke in the middle, leaving a six inch bit and part of the jars in the hole. They are being fished for but probably little will be accomplished until the arrival of the appropriate fishing tools. Two days work with them should be enough to determine whether the hole can be saved or whether it had better be abandoned and the derrick moved far enough for a new hole."

By this report of January 15, 1921, there is considerable lost metal in the hole. Several bits and jars and a casing shoe as well as the underreamer of course which is still upside down. It is easy to think that they should be able to get some of this equipment out of the hole with the fishing tools they have because this is a part of drilling of which they are familiar. While it is true that tools get lost on occasion, it is important to remember that some of these items weight hundreds of pound and keeping a strong positive grab on them is needed to pull them up over two hundred feet. There must have been a level of frustration that was off the scale. Holland commented on having lost the ability to be polite when fighting the gasoline engine before drilling began. One can only imagine the kind of words being used at the rig and the level of tension after days of trying to snag something blindly at the bottom of a hole 250 feet deep.

Mr. Holland sets an end for this hole of just a couple more days of fishing after the proper tools arrive. If the lost equipment cannot be raised or the hole cleared at least enough to continue drilling, he will declare it lost and move on to the next option for drilling on the rim of Meteor Crater.

There is finally some good news for Holland to report on January 22. "Immediately on the arrival of the bit hook it was put to work and it recovered the bit lost on January 10th after a few hours fishing. Strangely enough the bit and jars lost on the 15th were not recovered until the next day though they should have been on top. Nothing could be found of the lost under-reamer nor another lost bit. Evidently they have now been side tracked as the 8 ¼" casing went right to the bottom today, at 286 feet. The lost casing shoe has been about drilled up. Unfortunately the box broke on the only repaired stem so far returned from California, so drilling is being done with a light string of tools and progress is slow. The Union Tool Company promised to ship two stems on Monday night 24th January."

It would have been interesting to know if it was the repaired end which broke again or the other end that had not broken previously. But Holland does not note that detail. Clearly they should have just stopped a month earlier and ordered the heavier stems that Boston had agreed to use if the drill was moved to a new location. They should have begun using them at the current drill hole rather than continuing to break the ones they have. They say that hindsight is easy but in this case it should have been obvious to Holland that the stems they were continuing to use were defective. They had waited long periods already for other equipment such as the bit hook for fishing the lost tools.

They got most of the equipment out of the hole. Holland sounds confident that the other parts are out of the way and will cause no more problems. He is expecting that once the casing shoe is completely beaten up and passed that the hole will be clear. Once he gets a proper length drill stem of 30 feet, they can begin to make some faster progress. They are now quite close to the end of the limestone and likely still believe that the softer sandstone will be easier on the tools and easier to drill straight.

On the following day, January 23 Holland writes a letter to Jennings on an entirely different topic. It is a reply to a misdirected letter Jennings sent on January 12th. This letter is regarding sample collection. Holland reports that he ordered screw top containers from Union Paper Company, New York which are not as fragile as glass jars. However, he ends the letter with a mention that there is ample room on either side to move the rig five feet and ends by saying that "fortunately the necessity of moving the derrick has been avoided for the present." He expresses renewed confidence that they can continue drilling down.

There are a series of telegrams exchanged again during the week of January 22th to 27th. Barringer is copied on the weekly reports, but he does not receive them directly from Holland. He appears to get the information from Boston. He is often a few days late in his remarks, and that is the case here.

New York January 22, 1920
To L.F.S. Holland, Canyon Diablo, Arizona
"Barringer supports use of sulphuric acid poured down hole through rubber hose as a possible means of dissolving hard limestone nodules and possible releasing broken tools. Wire me Boston what you think of suggestions and also present condition of hole."
Sidney J. Jennings

Holland replys from Canyon Diablo on January 25th
Sidney J. Jennings 55 Congress Street, Boston, Mass.
"Your blue twenty second see no advantage in Barringer's scheme. Lost tools were

always loose and flopping around in big hole and crevices. Hole now 290' with great deal of iron in bailings. Expect much faster drilling on arrival of repaired stem on way."
L.F.S. Holland

Boston, Mass January 25
To L.F.S. Holland, Canyon Diablo, Arizona
"Your wire twenty fifth, if iron in bailing show test for nickel by using dimethyl gloxime then please have sample sent by parcel post to Mr. Clevenger 55 Congress Street, Boston for analysis."
Sidney J. Jennings

Canyon Diablo, Arizona January 27, 1921
To Sidney J. Jennings, Boston, Mass
"Tests show no trace. Hole in sandstone. Fishing job again."
L.F.S. Holland

Because of the strangeness and awkwardness of how the telegrams read it seems appropriate to mention again that they as with the quotes of Mr. Holland and others, have been reproduced here with all their errors, lack of punctuation, and The Western Union added bits. Over the course of a lifetime sending telegrams an operator might gain great skill at deciphering the handwriting of the American public. Still one could not expect them to spell dimethylglyoxime correctly as a single word.

On January 29, 1921, Holland and the crew have reached a milestone so to speak. They have made it through the limestone into the sandstone. They still have hopes that it will be easier going from here. But, Holland closed his telegram above by reporting they are in a fishing job again. "At 288 feet sand commenced to show in the bailing from the drill hole and at 290 feet the washed sludge was practically all brown and yellow stained saccharoidal sand with a great deal of iron and steel. The latter gave no reaction for nickel. The same condition is found at the present bottom of the hole 295 feet. It holds water fairly well."

"A repaired stem and sinker were shipped from California early in the week, but have not yet reached Sunshine, so drilling has had to be done with a light string of tools. To further retard drilling, one of the lost tools fell from behind the casing into the hole, in a flat position, on 26th instant. As the tool could not be fished out with the casing in, the casing has been pulled, and drilling by and fishing have continued since. If the tool can be gotten into the hole in a more or less upright position, it can be fished out."

"Since the series of calamities to the tools, there has always been the question whether it would be less unprofitable to abandon the hole and move the derrick or whether fishing and drilling by should be continued. The latter has always seemed the lesser of the two evils, especially as a new hole anywhere near the present one would no doubt have trouble with the crevices in the limestone which have been so detrimental in the present hole. Now the underlying sandstone has been reached, the drillers expect much easier going, even if the sandstone is fissured also."

The struggle continues with Crater Mining Company Drill Hole No.1 but Holland is still working through the problems, and nothing is yet an unsolvable problem. There is the fishing for the drill bit that has fallen out from behind the casing, and it must be removed. But there is still a chance that it will be easier soon. Holland's February 5th weekly report has that hopeful spirit about it. "On the arrival of a repaired thirty foot stem on 2nd instant it was immediately hitched on and "drilling by" continued. The hole is now 311 feet deep in sandstone which holds water. Drilling has been retarded by a piece of round steel which moves around in the bottom. During last night the lost 8¼ inch bit was gotten in an upright position in the hole and fished out. It shows few signs of having been battered but over a foot in length at the pin end is broken off, and no doubt this is the piece now in the bottom."

"The spud ordered on December 29th would be very helpful in getting this piece of steel out of the way, but has not yet arrived. If conditions permit, I will go to California in the next few days in the hope of getting some improvement in service for tools and repairs from the present people or someone elsewhere. The handicap now is greater than it should be, even in this isolated camp. Another matter that needs attention is freight. Last night a telegram arrived stating that a shipment of tools for repair, weighing twenty-six hundred pounds had been charged as a twenty four thousand pound carload by the railroad. Though a flat car was provided, the charge should be the actual weight, and I anticipate little trouble in adjusting this matter with the Santa Fe Railway officials at this end, but the Southern Pacific Railroad at the other end may be more difficult."

Holland sent a letter to Sidney Jennings on February 7, 1921, that was devoted to other topics than the problems with the drilling. Mostly a concern for the amount of water in the reservoirs for the drilling program. The hole has been holding water well enough for two hours of "drilling by" without refilling. But, the crater has had no significant rain that added to their supply. There has been little snow also. In the last paragraph of this letter, he briefly states the following. "The third lost bit was recovered on Friday night 4th. The under-reamer is now giving similar trouble, and the fact that it is upside down greatly adds to the difficulty of fishing for it."

Thus begins the darkest time of Holland's work at Meteor Crater. The exchange of telegrams about whether the hole is lost and what to do if it is. The discussion of digging a tunnel moves to the top of the list for methods to preserve the work already done, and for clearing the hole. Mr. Anderson is a major player in this second flurry of telegrams.

Los Angeles February 10, 1921
To Sidney J. Jennings, 120 Broadway, New York

"Under-reamer broken in two. No chance recovery. Obtained man with equipment for moving derrick. Address until Saturday 1718 La Brea Ave, Hollywood."
L.F.S. Holland

New York, February 11, 1921
To L.F.S. Holland
"Your wire tenth don't move derrick until A. P. Anderson has chance of investigating situation. Anderson due in Fierro, New Mexico, Saturday twelfth. Plans to be Crater Thursday seventeenth. Get in communication with him so as to be sure to meet at Crater. Reduce expenditures to minimum."
Sidney J. Jennings

Los Angeles, February 11, 1921
To Sidney J. Jennings, 120 Broadway, New York, NY
"Detained rigger until Anderson arrives. Drilling crew gets straight time."
L.F.S. Holland

Los Angeles, February 11, 1921
To A. P. Anderson, Hobart Building, San Francisco, Calif.
"Under-reamer cannot be recovered. Hole lost. Jennings wires not to move derrick until your arrival. Arranged with Wininger's man to leave here with jacking equipment Saturday morning but will stop him unless you wire contrary. Address 1718 La Brea Avenue, Hollywood."
L.F.S. Holland

Los Angeles, February 11, 1921
To A.P. Anderson, Fierro, New Mexico
"Under-reamer broken in two cannot be recovered. Secured rigger to move derrick but stopped him leaving on receipt Jennings wire not to move derrick until you arrive. Wininger's jacking equipment already expressed. Secured heavier string of tools in accordance Jennings previous approval. My address Saturday 1718 La Brea Avenue, Hollywood."
L.F.S. Holland
Copy of above to Mr. Jennings by mail.

San Francisco, February 12, 1921
To L.F.S. Holland
"Mr. Anderson en route Fierro. Expect visit Crater about ten days. Have transmitted your message to him at Deming."
L. S. Rankin

Deming, February 12, 1921
To L.F.S. Holland
"Wire me night letter Fierro New Mexico particulars your trouble Crater and plans you

have for future drilling. Expect leave here Wednesday evening for Winslow."
A. P. Anderson

Los Angeles, February 12, 1921
To A. P. Anderson
"Your wire received. Anticipating having to abandon hole Mr. Jennings said he preferred jacking derrick five feet toward Crater. Though close to rim now movement few feet forward feasible with some extra foundation timbers. Might move back as much as twenty feet without disturbing engine but hole less likely of results than forward. Loss of tools due mainly to drilling in fissured limestone with too light stems and joints. Heavy string being rushed factory for shipment Monday. When instructed rig builder sending only his foremen our crew helping. Unless instructions changed will meet number seven train Winslow Thursday and will ask rigger foreman to be on ground. Time important as drilling crew getting straight time."
L.F.S. Holland
Copy to Mr. Jennings by mail.

On February 15th, 1921 Mr. Holland makes a weekly report to Boston, but by then through the telegrams, the parties already know that the hole is lost. The report is of interest and contains the full story of the last week. "Drilling. On the arrival of a spud on Sunday 6th instant an attempt was made to get the lost under-reamer in such a position that it could be fished out in a similar manner to the bit recovered on 5th instant, as it was preventing drilling through having slipped from behind the casing partly into the hole. It was discovered, however on Wednesday 9th that the under-reamer was broken in two, and the two parts were lying flat and completely covering the bottom of the hole so that not even the spud could get by. The drillers who had hitherto been confident that the tool could be fished out or side tracked became satisfied that there was now no chance of saving the hole. I immediately got in touch with Mr. Wininger, the rig builder, and saw him personally in Los Angeles on Thursday. In accordance with the plan approved by Mr. Jennings to move the rig five feet north, towards the Crater. Mr. Wininger got the necessary jacking equipment together and expressed it to Canyon Diablo, and it is now at the Crater. His foreman, who was to have had the assistance of the regular Crater crew, was to have left Los Angeles with me on Saturday, but on receipt of telegraphic instructions on Friday from Mr. Jennings not to move the rig until Mr. Anderson's arrival, I postponed the foreman's departure until Thursday 17th, and he is expected to arrive at the Crater on Friday morning. The total depth of the lost hole is 312 feet."

In the remainder of this long weekly report, Holland also discusses the heavier tools he has obtained based on Mr. Jennings prior approval. They are four and one half inch stems and 3x4 joints. There was going to be a production time with Union Tool Company that Holland felt was too long, and he got an agreement from the Regan Forge and Machinery Company of San Pedro to give the order precedence over all other work in the shop. The tools were already on their way to Sunshine when he made the report just three or four days later. He remarks that their price was also a little lower than that of the other firm. Holland arranged with Regan Forge to do their repairs as well and to have the future materials sent to Los Angeles by truck to avoid the two railroads which had been the cause of some of the delays in the past.

The hole is lost, and the topic now turns to how to proceed. A. P. Anderson sends Jennings his analysis of the situation from Winslow by telegram on February 18th, saying there are four possibilities. First, moving the present derrick five feet north, second moving the present derrick twenty feet more or less south. The third choice make a new excavation and erect new derrick at an elevation three hundred feet lower and three hundred feet north. The fourth choice drive a tunnel from the crater wall to the bottom of the current hole, recover underreamer, then drill the present hole deeper. He goes systematically through all the four options and finally ends with this statement. "Unless there are conditions I am not familiar with which would influence this work I would strongly recommend driving tunnel, freeing present hole and continue same to depth. It is now through the limestone and should make good footage when cleaned of broken tools. Believe breakages have been due to fissured condition and hardness of limestone which caused unusual strains and crystallization of weakest parts."

Sidney Jennings replies on February 21st to Holland about Anderson's Western Union message. "Have received Anderson's night letter eighteenth. Will decide Wednesday whether we will quit or continue work. If we decide on doing anything it will be to drive tunnel recommended by Anderson. You can therefore discharge drillers and proceed with your survey of tunnel. Elevation of tunnel should be such that back of tunnel is few feet at least below bottom of hole."

Holland made a simple one-line telegram response from Canyon Diablo on February 22, 1921. "Your wire twenty first received and instructions followed."

L.F.S. Holland makes one more weekly report that contains information about the drilling. On February 21, 1921, he wrote to the men in Boston mostly what they already knew. "On receipt of telegraphic instructions from Mr. Jennings tonight, the drilling crew were paid off and they leave for California in the morning."

Holland adds at the end of the report "The survey of the tunnel site in the direction of the drill hole will be made immediately." Several days later on February 26th Jennings instructs him that the "elevation of mouth of tunnel should be at least twenty feet below bottom of hole." Holland replys to this telegram the same day with again a single line. "Starting tunnel floor twenty three feet below bottom of hole."

Several months will pass as the tunnel is dug. Holland will be gone when the tunnel is just halfway to the drill hole. When it is completed the underreamer, and the rest of the metal in the hole will be cleaned out. The plan was to begin again using the 12½ inch bit with the heavier stems and joints Holland had obtained. But there is information that indicates a rotary drill was employed on the restart. The depth when drilling begins again is 326 feet. A large opened area appears to have been created at the end of the tunnel since the drill hole bottom when Holland stopped was 312 feet.

Other than constant concerns about the amount of water being too little to restart drilling Holland's focus moves off the drilling and the rig and rests totally on the tunnel until the last day of April 1921, when his work is finished.

The Drill Log

The drill hole on the south rim of Meteor Crater remains a conundrum still today. The location of the mass claimed to have been reached is far distant from where the transient cavity formed and where the asteroid should have vaporized. The logs of this drill hole do not read as if a single large mass was hit regardless of how weathered it was. Much was a rubble pile with pieces of red sandstone and small shells even in the final fifty feet. There is no doubt that some of the material drilled through was meteoritic. The best nickel tests made were in the last 25 feet well past 1300 feet down. The material was hard and described using phrases like "very black" or "mostly black or brownish" and "very magnetic" this does sound like the weathered iron meteorite.

The drill logs were written by experts in drilling operations. Beginning the log is L. F. S. Holland a well-respected mining engineer. C. W. Plumb, a very qualified professional in the mining industry, continues the log to its end. They are both bad-mouthed by Barringer and his sons later, but they were both experienced mining engineers and mine managers.

Today, drilling holes to great depth can be made many different ways. Even a hundred years ago there were some options. Holes drilled into any soft, creviced or broken rock would be cased with a metal pipe. This was done at Meteor Crater. To accomplish putting down the casing the hole had to be bigger than the outer diameter of the couplers that connected the sections of pipe. The hole always had to be enlarged to the size of the casing couplers. There were double action type bits in use in the 1920s that could do both the drilling operation and the enlarging or underreaming of the hole. But often the hole was processed twice. The drilling bit would cut down for a distance and then be pulled up. The underreamer bit would be attached and run down the casing. As the name states, it reamed the hole to the larger size. The casing would then be driven down with another section of casing being screwed on at the surface. Then the drilling bit would go back down, and cutting continued. The actual string of tools could be thirty or more feet long. It was not just the actual bit but also long parts that added weight and assisted the pounding of the bit and the pulling of the bit back up after each blow on the rocks. Unlike the cable to which the tools were attached, the string of parts at the drill bit end was rigid and often brittle, made of forged iron and steel. It was essential that the hole cut down straight and true. Wandering and curving would inevitably result in the tools breaking off and having to be fished from the hole. After clearing out any lost tools and repairing or replacing the parts a crooked drill hole would be straightened. Then drilling down could proceed again. They might have to drill the same spot several times just to the side of the first try until the wandering portion was completely removed by widening. Casing pipe was employed to prevent material from caving into the hole from the sides and above. If boulders fell behind the drill bit, they could become wedged and prevent the drill bit and tools from being pulled back up to the surface. Exactly this finally happened to stop the work at a depth of 1,376 feet.

The native rocks of Arizona were hit by an asteroid traveling more than 25,000 miles per hour, weighing an estimated 150,000 tons. The asteroid excavated the crater more than three-quarters of a mile across in seconds and shattered the rocks in several ways. Some of the rock was pounded by this comic hammer blow into powder so fine it is not even gritty when placed between teeth. Other portions of the sandstone were melted and squeezed into chunks of hard vesicular glassy rock. The thick layers of limestone, dolomite, and sandstone were cracked and fractured forming voids and cervices. These crevices gave the drillers many difficulties.

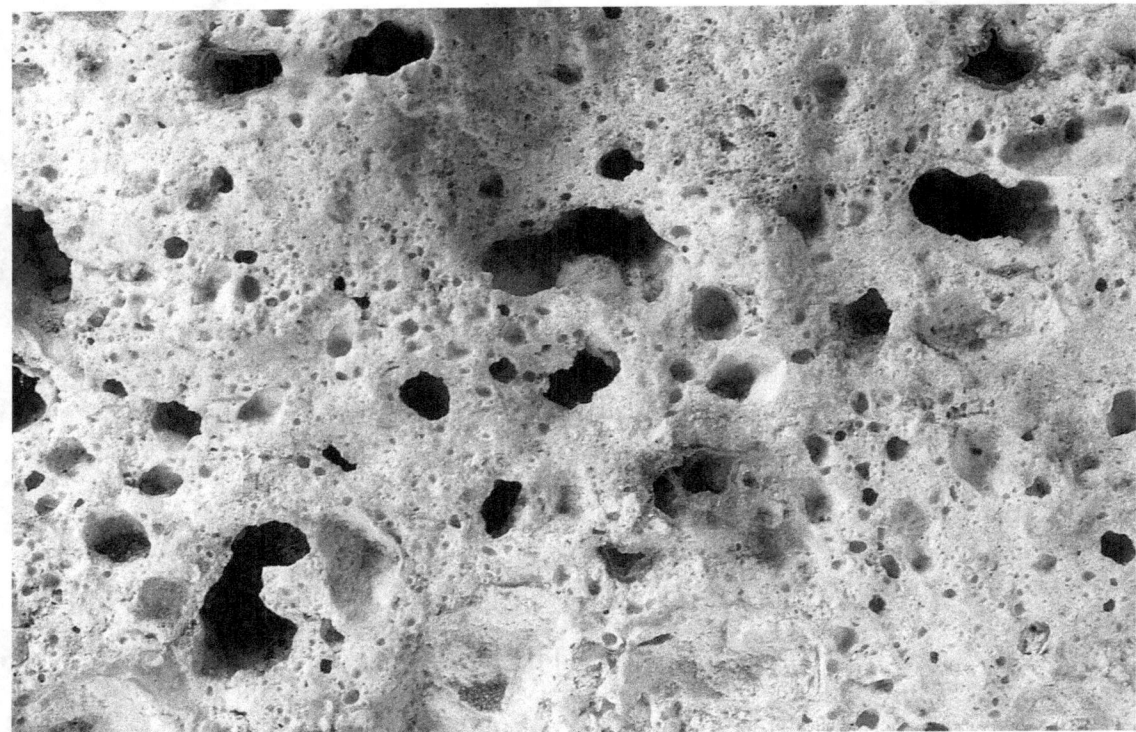

Shown here is a piece of what Barringer named Variety B of the metamorphosed sandstone. It has been partially melted and has become porous. Many of the vugs in the stone are lined with glass. Some of the voids are filled with threads of glass that crisscross the small cavities. This material is not very hard and the Variety A material is quite soft being sandstone that was shattered and squeezed receiving a sort of ghostly layered appearance.

It is astounding that the small diameter drills used on the crater floor in the first years of exploration hit obstructions with such regularity. The drill rig placement can be thought of as random because no one could know what was below the surface. With less than thirty holes scattered about the crater, the sampling was also small. Barringer restricted the drilling to the center so only about a twentieth of the crater floor was explored by drilling. To hit obstructions as often as they did there must be thousands of such meteorite fragments lying deep under the deposits now forming the crater floor. There may likewise be large numbers of meteorite fragments around the crater wall deep under the talus. Barringer was quite gratified when two shaleball meteorites were found plastered against the crater wall during the digging of the horizontal tunnel. They were found after the excavators had passed through the talus and were facing the layer of white sandstone. The shaleballs were blown there at the time the crater formed. They did not arrive there as a result of later erosion or landslide.

The log ends with the following statement by Superintendent Plumb. "Strong nickel test. Samples look as if we are passing through a recemented mass of conglomerate as we find shells rounded pebbles of red sandstone and of limestone, and also a great many small brown pieces resembling shale ball. Stuck and had to abandon at 1,376 feet. Bit appears to have wedged under boulders."

I would add here that Plumb has used the word boulders to describe rocks the size of baseballs in the entry for 1335-1339 feet. It sounds as if caving has occurred in the bottom of the hole. The last entry begins with the words "Extremely hard and rough" so it was no longer the soft mix of silica sand and sandstone but likely more of the hard iron shale nodules. This was the conclusion Barringer reported stating that they had drilled. . .

"through about thirty feet of undoubted meteoric material, that is to say, highly oxidized meteoric iron or so-called iron shale, cementing small fragments of sandstone and still smaller fragments of Variety A and Variety B of the metamorphosed sandstone previously described. The drill is immovably stuck at this depth, but so much iron shale has been brought up as to make the conclusion inescapable that the main mass of the meteorite is underneath the southern wall of the crater."

There is much in Barringer's statement that is questionable. The words used at the beginning of the quote should not be forgotten by the reader toward the end of the quote. It was never a continuous mass of meteoritic material but was always changing and always a mix of rock material. It is evident to most scientists now that no main mass survived the impact. Most of the meteorite was vaporized and became the ever-present tiny iron spherules found in the soil around the crater. But, there is evidence enough to show that many fragments of the mass were blasted through the rocks into what would immediately become the crater walls. Other fragments were shot up and out at the time of the impact and survived being vaporized. What percentage of the mass survived? Probably not too high a percentage but enough to constitute thousands of pieces buried under the floor of the crater, around its walls and out on the surrounding plains.

Though this book is about the drilling on the south rim, it seems appropriate to say a little more about the holes drilled into the crater floor. Several had obstructions of hard material which gave strong nickel tests and produced metal particles while being drilled. As on the rim heavy black residue along with green color was seen. Any water in the drill holes on the crater floor would quickly become green. Magnets recovered metal particles from some holes. The metal was not from the bit, rod, or casing. Dynamite failed to move an obstruction or crush it in one hole on the crater floor. Many of these pieces were likely fragments that were disrupted at the moment of impact and after being ejected nearly straight up fell back within the confines of the newly formed crater. The obstructions were hit at depths too shallow for them to be fragments blasted downward into the native rock. On only a couple occasions did drills on the floor succeed in going deep enough to reach the undisturbed rocks beneath the crater. Many of the holes put down into the floor had problems very similar to those experienced later at the south rim drill site.

A recent theory is that the nodules and hard nickel bearing rocks found under the south rim are not primary meteorite material, but are secondary deposits resulting from the impact. Nickel-iron liquid injected into fractures and cavities weathered to a nickeliferous mineral. This might explain the zone of mixed and alternating layers of hard iron shale like rocks. But would liquid injection explain the two distinct and isolated metal bearing objects that were nearly too hard to drill through?

Based on the final log entry it would appear that it was a simple cutting type bit stuck at 1,376 feet. But in the case of the breakdown at 312 feet, it was reported as the lost underreamer bit. Holland was very clear that the underreamer was broken into two pieces which were lying flat across the bottom of the hole. It is stated that the casing was lying on the underreamer too. Today we might send down a video or fiber optic camera to see what was going on. Today we might be able to grab the lost tools with the assistance of a camera. But in 1921 the men were 312 feet up on the other end of a small diameter metal tube and working blind except for impressions they could take of what was at the bottom of the hole. They could send down a block of soft lead on the end of the drill string and bang it into whatever was at the bottom. Inspecting the way the impression block was dented would give them a clue as to the extent of the blockage and maybe the nature of the item causing the blockage.

Many times during the drilling tools and obstructions had to be bypassed. The 312-foot obstruction, however, forced the digging of the horizontal tunnel into the south wall. This was the depth when C. W. Plumb began his tenure as superintendent. The log records that drilling began again at 326 feet in depth. Did the blockage by the underreamer at 312 feet cost L. F. S. Holland his job? He had a prominent career in mining all over the Western United States also in Nova Scotia. His name is often seen in mining journals, and his career began long before his time at Meteor Crater. He was asked in a communication from Sidney Jennings the Vice President of US Smelting to resign. He was told that they are shutting down some of their exploration projects. To take care of employees who had been with the company much longer, he was being replaced by a man from one of the closing properties. Holland resigned as of May 1, 1921. There is no doubt that the drill hole had been a nightmare. If by removing Holland they hoped someone new could do better that hope was never realized.

L. F. S. HOLLAND
SUPERINTENDENT

CONSIGN FREIGHT PREPAID
TO SUNSHINE, ARIZONA

THE CRATER MINING COMPANY
CRATER, VIA WINSLOW, ARIZONA

April 13th. 1921.

Sidney J. Jennings Esq.
Vice President,
United States Smelting Refining & Mining Company.
822 Kearns Building,
Salt Lake City, Utah.

My dear Mr. Jennings:-

I have your letter of April 7th. and in accordance with your wishes, beg to resign as from May 1st. next. I shall be very glad to do all I can to assist my successor here in getting started right.

While I regret very much the occasion for this resignation, I hope that a change in conditions may lead to a renewal of our association somewhere. It may be that you or Mr. Anderson or Mr. Moore can give me some of your field work at times. There are few men who have had a more varied experience in mine examination work than I have.

In the meantime, if you can give me such a letter as will help me to find other employment, as you kindly suggest, in these rather desperate times for mining, I shall be grateful. My address will be as before:- 1718 La Brea Avenue, Hollywood, California.

Very truly yours,

It seems too easy to believe that incompetence led to all the troubles and ultimately the lost underreamer blocking the bottom of the hole. Holland's years of prior work seem to rule out a lack of experience as well. His weekly reports show much cleverness, ingenuity, and knowledge. It is more likely considering the continuing disaster that this hole proved to be that he was just present at the first and worse of the problems. C. W. Plumb his replacement would experience a similar event in just days and only 58 feet deeper. The drill twisted off again at 384 feet. Later objects fell against the tools, and more tools were lost in the hole, and more obstructions had to be bypassed. All just the same troubles Holland experienced. All were just as difficult for Plumb to work through as they had been for Holland. Reading some of Mr. Holland's published magazine articles reveals he was a man with vast geological knowledge and well versed in chemistry. The wide range of equipment discussed in other articles also reveals that he was mechanically adept.

The crater presented the worse possible set of conditions for churn drilling. They began with a 10-inch bit and tools. Then used smaller and smaller diameters as they had to bypass lost equipment or sort out other problems. Later after the situation became serious, Holland suggested, and the bosses in Boston agreed that a 12-inch diameter tool string should be used once the hole was cleared and the casing fully pulled out. The 10-inch bit could not cut a shoulder into the steeply angled fissures to continue going down straight. Instead, the 10-inch bit ran off down the slope of the fissures and got crooked causing the brittle iron to break. The hope was that the 12 inch would be large enough to stay true as it pounded down. The smaller diameter bits and tool strings he was forced to use were even more likely to follow the fissures and break. Holland's drill string of choice when he had it was thirty-feet long. It consisted of several parts attached to each other. The whole string though would need to stay straight, or it could break dropping off hundreds or even thousands of pounds of parts in the hole. Anything lost by a breakage would need to be retrieved using "fishing tools" with which every driller was very familiar. If the lost tools or other parts could not be fished out, then they had to be bypassed when possible. To determine this an impression block made of soft lead was dropped on the obstruction. When the impression block was pulled up and inspected an imprint of the obstruction would often show. If the hole was only partially blocked, a smaller diameter drilling string could be sent down so drilling could continue. Losing equipment was not uncommon and having to drill past obstructions was not uncommon either. But once you are down to 8" tools in fissured rock breakage becomes harder to avoid. Another cause of problems was the original equipment supplied with the rig. The drill stems were much lighter and thinner than replacements Holland ordered. He obtained heavier drill stems commonly used in California drilling. The four stems that broke in a single week all failed near the end where the boxes were. It was suggested that the metal crystallized from the pounding and this caused them to break when even slightly bent. But the breakages all being at the same place on the stems whether 30 feet long or 18 feet long or 10 feet long seems to support that they were too light for the work and poorly designed and improperly manufactured. This idea is further supported by the fact that they broke on several occasion after only a short period of use. Not really enough time for them to have suffered from the theorized metal fatigue.

Mr. Holland is spoken of in very negative terms later in a paper written by Brandon

Barringer in 1964. Brandon's comment is "one inexperienced superintendent replaced with another" referring to C. W. Plumb. But both his father and brother Reau had a cordial relationship with Holland while he worked at the crater. Reau is often mentioned in the letters that D. M. Barringer wrote to Holland. Reau offers his best wishes also to the superintendent whenever he is present at the time a letter is written. D. M. Barringer himself used harsh words for Holland and his replacement saying that U. S. Smelting is using "poor drillers" and "poor steel." But Holland reported to very qualified miners at the home office in Boston. Many of the early reports from the Summer and Autumn of 1920 were addressed to C. F. Moore and A. P. Anderson. These two individuals worked with each other for years at several mines doing the same work as Mr. Holland. There is never anything in their correspondence that implies a lack of quality in Holland's work. Holland's later reports are copied to these men but sent to Sidney J. Jennings who had one of the most distinguished careers in mining imaginable, and he never mentions to anyone that Holland is unqualified.

In the same 1964, Brandon Barringer biography of his father the younger son reports that the drill rig was changed from a churn drill to a rotary drill after the clearing of the hole and the restart with Mr. Plumb. This change seems not to have resulted in fewer problems. It would have added considerably to the expenses of United States Smelting Refining and Mining Company as well.

Considering the range of responsibilities placed on Holland he seems to have stood up well during his time at the crater. He was expected to be host to visitors, shipping specialist, purchasing manager, record keeper, personnel manager, mining engineer, and overall supervisor of the entire operation. He did the banking and the payroll. During his time at the crater magnetic and scientific studies were conducted. Plans and drawing of the crater were produced. Holland was responsible for coordinating these activities and getting the information to those that needed it. Just consider that the pipeline from the dams and reservoirs was approximately 15,000 feet long and was completed in just several months. The laborers many of whom quit each payday had to be replaced by Holland. Much blasting was done under his supervision. Workers were injured, and others got sick, and he had to take responsibility for getting them help. These may have all been individually routine activities, but the sheer number of tasks he had to perform daily was remarkable. No one likes to have several bosses, but that is what Holland had. He was an employee of Crater Mining Co. itself a subsidiary of U. S. Smelting Refining and Mining Exploration Co., and he reported to several persons there. Barringer kept his hand in the business and Holland was expected to satisfy his requests too. Barringer's sons were also involved and visited the crater as did one of Barringer's brothers-in-law from Phoenix. Holland regularly sent information, or samples of rock and meteoritic material to individuals whenever Barringer asked. He was expected to track the packages as we would say today. He handled the shipment of ½ ton sample of silica sand from the south slope to the McKnight Firebrick Company in Porterville, California. That job for Barringer was not related at all to the work of drilling on the south rim, but if it had been a successful venture USSR&M. Co as the leaseholders would have been in charge. Nearly all of these activities for Barringer required numerous letters and telegrams to arrange them and to report their progress. Telegrams were sent and received at the Western Union Office, 208 Kingsley Ave. Winslow, Arizona over 20 miles from the crater. This required a trip over dreadful dirt roads. Some telegrams were delivered by a part-time man who brought them to the crater for a fee. But others required a trip to Winslow using the car or truck Crater Mining Co. owned. Today the corner of Kingsley Ave. and Route 66 is the "Standing on the Corner Monument" in historic Old Town Winslow. A quick drive down I-40 now, but in 1920 it was a long hard trip from the crater. A trip impossible to make after a rainstorm. Holland was a man with too many tasks and too many bosses, trying to do an impossible job under poor conditions.

Silicon carbide and diamond drilling bits used today cannot be compared with the churn drilling rig used at Meteor Crater on the south rim. The bit pounded up and down at the bottom of the hole and the tremendous weight in the tool string crushed and pulverized the rock. Meteor Crater had some soft rocks and some hard rocks. The native rocks of the area would have been no big problem even in the 1920s. Just flat solid layers of limestones, dolomites, and sandstones. After the impact, the rocks at the crater became a different story. Asteroid fragments were propelled into the subsurface all around Meteor Crater. Add to those impenetrable rocks, the fractured rock layers, and the cavernous voids underground, now you have a driller's nightmare in the 1920s.

Another company formed later would drill holes in the same area after Crater Mining Company Hole No.1, but none of those were to a similar depth. U. S. Smelting, Refining, and Mining Exploration Company decided there was no value in the deposits under the south rim of Meteor Crater. They felt the metals had been leached out during the process of weathering to iron shale. Soon after the drilling, the parent company in Boston gave up their rights to the property. Barringer was again on his own and in control. The cost to do this work on the south rim exceeded $175,000. Many holes were drilled much later around the crater perimeter by the U. S. Geological Survey. None of these were deep either. Later still seismic studies with explosives were done which raised questions about how the meteorite materials got into the rocks under the south rim. It seems there is never an end to what Meteor Crater can reveal to investigators or the new questions it can raise.

The Tunnel

March 8, 1921 - It has been just ten days since Jennings told me to drive a tunnel into the side of the crater and clear the drill hole of the lost underreamer. I hate the whole idea of the excavation and getting it started has been another huge amount of work for me. But there was not another answer. Mr. Fay left. He finished his survey on the 6th and put a stick in the ground and told me the bearing and the length. I think he will be one of the few people to make any money here at the crater. He reminds me of the general store owners in California during the 1850s. They sold all the supplies to the miners. But most of the miners never found more than just enough gold to buy more food and supplies. The store owners made the money. Mr. Fay has been here for days or even weeks at a time making surveys and magnetic studies. I have not gotten the bill for his latest work, but I know when I totaled his charges in October he had already been paid $1300 this year. And he was here for a couple of weeks after that finishing the new magnetic variation study. And now the survey to locate this tunnel. The company probably owes him at least that much again.

Mr. Fay's spot to start the tunnel is on a steep part of the south talus. We finished the trail down to his marker. It will take about 15 minutes for the men to hike from the tunnel back to camp. That does not sound like much of a walk, but after eight hours working in the tunnel, the climb out of the crater will be difficult for exhausted men to make. It is another switchback trail; only a mountain goat would try to climb straight up and out of Meteor Crater.

I had to get men from Jerome; there are no hard rock miners nearby. The three men arrived and started digging. They are in eighteen feet and are timbering the tunnel as they go. There are so many boulders if it keeps up the same we are going to use a huge amount of explosive on this tunnel. It was a dangerous idea, to consider. I hope no one gets killed doing this. We have had some close calls since I came. The first drilling foreman had that paralysis attack. Then that maniac with the rifle could have killed someone. Accidents and illness are always on my mind we are so far from a doctor. We need to do something about storing the explosives. The last thing we need is a case of dynamite going off on its own because of a rock slide or something else. Later in the year, the hot weather could make the dynamite sweat nitroglycerine. April 18, 1921 - I was in the tunnel most of today. The men have been working straight through for three days. The fine white sand just keeps running like water into the tunnel from above. We have made little progress since last week. We are in the most broken rock so far. The timbers that we salvaged from the crater floor which I thought too heavy for this job maybe saved a couple of lives. I would have never believed that 4 by 6 inch spiling would crack or the 8 x 8 and 10 x 10 timbers would either, but they have. The men got out, and we shored it up and have pushed in a little farther, but the sand did not stop until this afternoon. Almost suddenly the vast supply ran out. I held a lantern up into the crevice above us, and all I can say is I felt awestruck. Like being in the cathedral as a youth in London when the pipe organ was playing Beethoven and shivers ran down my spine. The top of the crevice was at least twenty feet up and barely visible in the lantern light. The narrow sides of broken rock were arching upward and met directly over the breast of the tunnel. We mucked out the remaining sand, and it was nearly sundown when I emerged from the tunnel and saw myself in the open. Covered in white powder, I looked more like a baker after an accident with a sack of flour than a mining engineer. Some of the men were really scared, and I think we have been very fortunate that no one was killed. We should be near the end of this part of the tunnel and sandstone in place should be hit in just a few feet. It will be safer I hope from here on.

After the big white spot in the center of Meteor Crater, the next most visible remaining evidence of the work done a hundred years ago are the two tailing piles on the south wall of the crater. One was an adit which does not show in a photograph taken when the real tunnel was already 174 feet long. It has been said that it was dug for storage. When completed at a length of slightly under 400-feet the real tunnel was used to reach the underreamer tools blocking the hole on the south rim.

The talus of Meteor Crater is the debris zone above the crater floor going up the walls to the point where the original layers of rock are exposed. It was into this boulder-rich rubble that the tunnel was begun. The miners had to drill and blast the boulders and then haul out the material created and dump it. Along with the large blocks was sand and smaller debris.

The shattered condition of the rocks was well known to D.M. Barringer, and he had cautioned the men of Boston about the dangers of digging the tunnel. But, there was little else left they could do. They had several choices, all bad from which to choose. First, they could move the drill rig on the south rim about five to ten feet and drill a new hole. This could be done with little difficulty using jacks, pry bars and manpower. The problem was the high likelihood that they would drill down and encounter the same crevices and sand runs that had plagued the original location. They might even punch into the old hole since it had been so crooked in places and required so much straightening.

The second choice was to move the rig 20 feet or more still staying close to the crater edge. The problem with this choice was that the site created for the derrick would not allow for a big move. The drilling site would have to be enlarged and leveled. Likely thousands of tons of additional rock would have to be scraped and blasted. They had removed more than 5,000 tons of rock to make the current 100' x 40' site. The cellar would have to be filled, and a new one blasted, and the rig was much more difficult to move that larger distance. The connection for power to the gas engine could be made for a short distance change. But, the engine was mounted on a concrete foundation. The engine would have to be moved and remounted. The drill rig was enclosed in a building, and its fully assembled weight was very great. Usually, the engine was surrounded by a building as well. There was a forge house but it is never stated if the engine was in a building also, but given the often inclement weather it is likely. They might have been able to keep the forge where it was and haul the bits requiring dressing further. But, the engine house and foundation would have to be rebuilt in a 20-foot move.

The third choice for the officers of the company in Boston was to abandon the drill site altogether and begin again at another location nearby. They could disassemble the derrick and rig and reassemble it after the new location was prepared. There was material on site for a second rig which they had expected to use. Those materials though incomplete because of the shortages from the supplier could supplement the materials if needed during the reassembly of the rig at the new location. This choice would be a very expensive both in money and time. It had taken months to clear and prepare the original drill site. Most importantly there was again the high probability that the rocks beneath any new location would be just as shattered, creviced, and sandy as those of the current location.

These choices all being bad, costly, or time-consuming left the dangerous choice of digging a tunnel to clear the obstruction from the hole they already had. There was the constant hope of the drillers that if they could get through the limestone and into the sandstone that the drilling would be easier. The drill had just reached that goal when it became blocked by the broken underreamer bit. They were down more than 300 feet and after all the problems that effort and money were something to try and preserve.

Crater Mining Company had a huge financial investment in the drill site at this point. It had cost $5263.69 to prepare the site as of October 15, 1920, two weeks before drilling began. It had cost $1190 for the rig builder's services. This was a cost that could be expected to be significantly higher if they had to disassemble and then reassemble the rig. The drillers would be dismissed and have to be brought back, a travel expense that had been $240 the first time, one way probably by the train. The water system and pipelines would have to be modified if the drill rig was moved very far. The water pressure for the camp and the rig was determined by how high the 13,000-gallon water tank was above them. Any new derrick location would need to be the same elevation. Otherwise, the tank would have to move as well or be raised at its current location. They had steel towers from tanks Standard Iron Company had used before. Holland asked D. M. Barringer how he felt about the request made by the government to take a tower and use it at one of their sites. Barringer's answer was that all the materials at the crater were at the disposal of the newly formed company and what was decided by U. S. Smelting was OK with him. However, he added that the towers might come in handy later so giving both away might not be wise. So the tank could have been raised with little expense since one of the tank towers at least was still down at the dam.

The tunnel choice was discussed for about two weeks, and on February 21, 1921, Mr. Holland confirms that he is going to begin a survey for a tunnel in the direction of the drill hole. Five days later he receives a communication from Sidney Jennings. It had been absolutely decided to drive a tunnel into the side of the crater. The elevation of the mouth of the tunnel was to be at least 20 feet below the bottom of the drill hole. The same day as that wire from Jennings, Holland replies that they will start the tunnel 23 feet below the bottom of the hole.

On March 2nd the survey has been completed by Mr. Fay and checked. They are going to tunnel in from west of the so-called MN line marked on the old magnetic survey. This location was selected to avoid following a "draw" where the talus was probably much deeper than on the ridge they did follow.

In just a few days Mr. Holland's men completed the trail to the tunnel site. He moved tools, rails and a car for hauling out the rock to the tunnel mouth. Mr. Holland acquired men from the area of Jerome, Arizona where there were hard rock miners to do the work. Holland asked an old friend who was the underground foreman at the huge United Verde Mine to pick three miners. They arrived on the property March 5th and began working the same afternoon on the tunnel. Holland is hopeful at first that the men can be hired on a contract basis. But, the unknown character of the rock may make that difficult. While the talus may be easy to tunnel through the limestone and sandstone may not, and the men will not want the same pay for working the hard rock as the talus rubble.

Three days later Holland reported to the officers in Boston that the tunnel was timbered to eighteen feet from the starting point. The miners had been working chiefly in large boulders of limestone requiring drilling and blasting. In the same report, Holland gives the exact information of the survey to the home office. The tunnel would be 390.66 feet long. The bearing of the tunnel is S 20 ° 46' E. The portal of the tunnel was dropped to 23.45 feet below the bottom of the drill hole. Holland will send cross section prints to Boston and Barringer as soon as they can be obtained.

The weekly reports by Holland show both the progress of the tunnel digging and the difficulties that are encountered. Problems we might expect today with the blessing of a hundred years of hindsight and many other craters to use for reference. The material they dug through is immediately a mixture of half large boulders and half fine sand. The two ends one might say of the pulverizing process of crater formation. Holland reports that the fine sand runs like water. At times it was not an easy matter to keep the tunnel from being swamped with sand. On March 14th he reports that he has ordered respirators because the fine dust is "of course, unpleasant for the men."

All the timbers for the tunnel have been salvaged from the old buildings in the crater. He states that it will only be necessary to obtain wood for spiling since none suitable for that remains in the crater. The tunnel is completely lined with wood where needed. Large posts and overhead cross members were made from the wood cannibalized from the old buildings on the crater floor. But, the thinner planks that filled the space between the large posts and beams called spiling were no longer available from the crater buildings. The spiling was placed over the beams and then driven forward with hammers while more pieces were put in behind the first and also driven in. The planks lap over each other as you dig the tunnel inward.

Even a small size tunnel like the one at Meteor Crater could use an enormous amount of timber and spiling in 390 feet of length. However, not the whole length was timbered. Once the tunnel was into the uplifted and cracked but stronger sandstone of the crater wall, it was timbered only as necessary. In one of his reports, Holland writes that two thousand feet of spiling were obtained and that they were still using the large heavier than needed wood from the crater buildings for the posts and crossbeams.

The fine sand or really dust of Meteor Crater was a problem for the worker during the clearing and leveling of the rig site and continued to be a problem for the miners in the tunnel. It had been a problem for the Standard Iron Company workers years before. Barringer writes that when they were working on Cut No. 10, the fine sharply angular dust had caused some of the workers to begin spitting up blood. Holland had gotten the tunnel miners respirators made by Goodyear Tire and Rubber Company of California. They had a sponge to filter out particles, and he said they worked well. But, the miners chose not to use the masks. Apparently, it was unmanly to wear them, and of course, we know it was very dangerous to work without them. On March 29th Holland was satisfied with the respirators and reports that he does not think it will be necessary to run a water line to the tunnel to spray down the dust. Just five days later in his Superintendent's Weekly Report dated April 4, 1921, the tunnel is at 144 feet. He reports the following about one of the miners who quit the day before "on account of sore lungs and I expect to have to lay off another for his own good as soon as a substitute arrives. Respirators have been provided, but the men leave them off most of the time. Water to help lay down the dust has to be used in moderation to avoid caving the ground."

By the following week of the 11th of April, they have also added a blower and air pipe in the tunnel to improve the air conditions. Another man had to be laid off that day because of bad lungs. This is likely the man mentioned in the previous week's report. There was now just a single miner on each shift and muckers who were actually cowboys. The face of the tunnel is in 174 feet. They are in the white sandstone and not all of the tunnel needed to be timbered beyond this point. However, some of the sandstone was so shattered that it caused caving for days. They expected the rock would need less timbering as they progressed further. By the following week they were in just 186 feet and that hope also quickly reversed. The ground became "particularly loose and difficult . . . sometimes caving to a height of twenty feet." That is just a remarkable statement to read. To be digging a tunnel about 6 feet by 6 feet in width and height and to suddenly have it cave away and run sand until there is an open cavity with a roof 20 feet above. There was at this time in the digging hardly any rock at the working face of the tunnel, which is called the breast. None of the material was larger than an orange. There was a near disaster from the sound of it, though he writes of it rather routinely for his bosses. Holland reports that 8 by 8-inch timbers and 4 inch by 6 inch spiling have been cracked by material running into the tunnel from above. It has been necessary to shovel steady for three days. The men "look as if they were working in a flour mill." During the digging, they suddenly reached a spot where a tremendous amount of loose sand came down on top of them, and it struck the tunnel timbers and spiling cracking them. Had they broken completely men could have been trapped or killed. Drowned as it were in the fine sand.

Miners often talk about sounds they hear in the depths of tunnels while digging. Some workings produce noises that give the mines a haunted reputation. Sounds of cracks and pops and creaks are common in mines. At Meteor Crater distant thundering sounds could be heard at the breast. Strange sounds whenever the spiling lumber was lowered over the cliffs at the derrick 200 feet away. Eerie sounds passing through the rock were heard.

Mr. Holland was replaced at the crater effective May 1, 1921. In his final weekly report on April 24, the tunnel is timbered and accessible to 203 feet from the starting point. That is where it has been for several days. All the footage gained since reaching that point has been lost to repeated big runs of sand. Every worker at the crater except the cook has been shoveling sand for three days. By putting 10 by 10 inch sets closer together and continuing to use the 4 inch by 6 inch spiling they hope that the flood of sand can be checked soon. He writes that "some of the men have been badly scared, but there has been no letup in the work." It is difficult to imagine working in a tiny space with sand pouring in so fast that for days you must shovel as fast as you can. The men must have worried about being covered alive by sand if the timbers and spiling failed.

On the morning of Saturday, April 23, 1921, in the heaviest winds of the season, the old shafthouse on the floor of the crater blew down. The boiler room annex which was next to it and a small portion of the annex side remained standing. For some time they had been removing wood from the old buildings to timber the tunnel. Whether the shafthouse was one of those building is not told. We are told that they had been sparing it from salvage a few months before. Mr. Holland is a remarkably lucky man. With but a week to go in his tenure at the crater he avoids being killed by mere hours. He and a miner were in the old Barringer main shaft the previous afternoon checking the water conditions. If the building around and above them had collapsed while they were inside it or in the shaft, they could have been killed. The only actual fatality of the collapse was one sheep.

There had been a drought at the crater for many months. The last rain to benefit the reservoirs had been the previous August. There was water to finish the tunnel digging operation but not enough to restart the drilling operation once the tunnel was completed. Holland had reported several times the low level of the water. Later on in his last report after describing the collapse of the shafthouse, he reports that just two inches of stinking water remained in the bottom of the shaft. He says that he was surprised to have found any water. The local rancher, a Mr. Hart, was a constant supporter of the efforts at the crater. He loaned men and materials and supplied tools and equipment. It was his mules that raised and lowered the timbers and spiling in and out of the crater. The water shortage near the crater was so great that he had moved his cattle to a different location where there was water. He was rounding up his horses to move them near the end of Mr. Holland's employment, during the tunnel digging.

L.F.S. Holland attaches to his final week report a log of the drill hole on the south rim. The last log entry is dated February 9, 1921, stating that the drilling foreman says there is no chance to drill past or around the broken underreamer lying on the bottom of the hole, and completely blocking it. In just two weeks the tunnel was begun. It was intact as of around 1990 when this writer spoke with a guide at Meteor Crater who had recently been in the tunnel to its end. But, by the time this author stood at its entrance and explored the area personally in 2001 it had collapsed somewhere beyond where you can illuminate it from the entrance. They were still fully timbering the tunnel at the beginning, but by 174 feet they were timbering only as they felt it was needed. It is possible that it is one of those untimbered spots that collapsed. Mr. Holland referred to the lumber retrieved from the crater buildings and used on the tunnel as being "generally heavier than one would choose." Still, a timbered part of the tunnel could certainly have failed after nine decades also. The talus zone is little more than a great many landslides piled up, with more waiting to happen. Perhaps one of the great runs of sand that the miners pushed their way through finally crushed the spiling and timbers. When saturated by rainwater flowing down the cracks in the crater wall such tremendous amounts of sand could become enormously heavy.

There is no sign of the rails or any evidence that an ore car ever ran through the tunnel. The floor of the tunnel is uneven as far as can be seen with a flashlight. Flash photography will illuminate to where the miners were after two weeks of digging. That was March 14, 1921, about forty-three feet from the entrance. The beams and spiling are still as they were in 1921. The wood is not gray and weathered like the scattered timbers outside. It looks fresh and strong. The second half of the tunnel was excavated after Mr. Holland left presumably under the supervision of Mr. Plumb. There is no reason to think he did not assume the same responsibilities at the crater which Holland had been given.

During the digging of the tunnel, D.M. Barringer was keenly interested in what the miners might find. As it happens, meteoric material was found just where the talus stopped, and where the in situ but tilted layers of sandstone began. Two shaleball meteorites were found at eight-one feet into the tunnel and fifty-eight feet vertically below the surface. They were about the size of a coconut each. Holland reported that just two feet of digging later the rocks were "in-place fractured white sandstone." Mr. Barringer would write that these shaleballs were "plastered" against the rocks of the crater wall. They were blasted there at the time of the impact and had stayed covered in the talus. Fifty thousand years they remained in place, decomposing from metallic iron into hard iron oxide until found by the miners. Barringer was thrilled to learn the miners had recovered the two weathered meteorites. Along with these two stones, the miners found the various other rocks that Barringer had discovered and described. Pieces of what he named Variety B; the vesicular metamorphosed sandstone had been dug out before reaching 71 feet.

At 203 feet of tunnel length, Mr. Holland is done working at Meteor Crater and leaves. His Superintendent's Weekly Report of April 24th with the attached drill log is nearly the last official paper that he kept from his employment with United States Smelting Refining and Mining Exploration Co. He had been the man in charge of their subsidiary named Crater Mining Company since its formation. His efforts never reached meteorite material buried under the south rim of Meteor Crater. He faced almost daily challenges and successfully began digging a tunnel into the side of the Earth's best-preserved impact crater. Certainly a notable achievement on its own and one of the fascinating features visible at the crater today.

Crater, Arizona Post Office

We get upset if we drive out of town and lose our cell phone service. We expect that we will always be able to make a call. In 1920 at Meteor Crater the telegraph and the United States Postal Service were the communication lifelines for Mr. Holland. He was receiving letters and wires, often several times a week. Ray Gebhart worked for Holland doing many different jobs. He was even Mr. Fay's helper during the magnetic survey, but most of the time he was the driver who came to the crater from Winslow. Gebhart brought out visitors on occasion, but he brought the mail and telegrams on a regular schedule. And he took correspondence back to the Western Union and US Post Offices in Winslow if Holland was not going into town. Crater Mining Company petitioned the federal government for a post office closer to the crater at Sunshine Station.

The first mention of this request for a closer post office was in Holland's weekly report dated October 23, 1920, when he writes the following. "As it now costs approximately ten dollars to send in the mail to Winslow, I have taken every possible step to get an office established at Sunshine. The name of Crater has been approved by the Postmaster General -there are already three other Sunshines- and the services of a willing postmistress secured at Sunshine, and the usual formalities have all gone through. I hope that the post office will mail every day instead of about twice a week as at present, and save the very unpleasant trip to Winslow besides."

That would seem to be a good start. That it was now just a matter of Holland waiting for the final approval to come through so that the mail would be delivered to Sunshine. Holland was British but had been in the United States long enough to know that nothing involving our government goes smoothly. Very soon his plans are turned upside down and backward by the Indian Trading Post at the little town of Canyon Diablo. Holland reports on this November 6, 1920.

"Application for a post office at Canyon Diablo was made by the Indian Trading Post there some months ago, and I understand that it will be open for business on November 8th. The Indian school already had a post office at Leupp. From a letter received tonight from the Post Office Inspector I gather that the Canyon Diablo people have misled the Inspector into the belief, which I am hastening to correct that the Crater Mining Company desires a post office there, though it would be far less convenient than an office at Sunshine (to be called Crater P.O.). We must frequently send to Sunshine for freight, especially gasoline, so that the carrying of mail would entail little to no expense, whereas it would generally be necessary to make a special trip to Canyon Diablo. Some of the 11½ mile road is in a very bad condition at all times, and is especially hard on tires. All three roads to Winslow are now very bad."

The background on the Indian Trading Post, which was really named Volz Trading Company was that it had been there for decades. They were collecting meteorites and selling them worldwide from the 1890's on until there was such a glut on the market that no one would buy more. Fred W. Volz had hired men to hunt for meteorites all around the area. He admitted later that he had sold approximately 10,000 pounds of iron meteorites. By 1903 when Barringer acquired the property of the crater's interior and began hunting for meteorites himself on the plains there were few left on the surface to find. By 1891 the locals working for Volz had already found the 625, 506, and 145-pound specimens and hundreds of smaller meteorites. Volz would later have the men walk and poke iron rods into the ground to find buried meteorites. Though the current manager in 1920 a Mr. H. W. Smith at the "Indian Trading Post" seems friendly enough from the few correspondences Holland has with him; it is clear that he is trying as we might say today, "to work an angle." He is always offering something that will give him a chance to make money off the Crater Mining Company. Holland is having none of it and is polite but does no "big deals" with Mr. Smith. D. M. Barringer will make mention of Mr. Smith in a few of his letters to Holland and had a good relationship with him.

Mr. Holland will not return to the topic of the post office until his weekly report of January 1, 1921. This report indicates that progress has been made, they are close to getting an official Crater Post Office.

"The postal authorities have ordered for the first three months a pouch be made up for us in Winslow and sent by carrier to Sunshine, now to be called Crater Post Office. The arrangement is expected to commence in a few days. The postmaster at Winslow has undertaken to send any Crater mail that may come addressed to Winslow, but it may be safer to address mail to "Crater via Winslow." Telegrams and express had best go to Canyon Diablo, as it is just half the distance of that to Winslow."

It sounds as if the Postal Service has put the Crater Post Office on a three-month trial basis to see how it goes. The specifics of how the pouch will get to Sunshine from Winslow are not described in this report by Mr. Holland. The wheels of the government grind slowly, but they do grind fine. By Holland's report three weeks later on January 22, 1921, the missing details have been realized, and the problems with the original plan are being worked out. Here is what Holland writes on that date.

"Although everything appears to be in order for the opening of the post office at Sunshine (to be called Crater) including the appointment of a postmistress and the provision of a pouch etc. there has been a delay through a mistake on the part of some official who overlooked the fact that Sunshine is on the railroad, and the postmistress lives alongside of the track, as he instructed her to hire a carrier to bring the mail from Winslow by road at a compensation not to exceed two-thirds of the stamp cancellations. Of course, nobody would carry the mail by road for such small pay, nor is there any necessity, as the Postmaster at Winslow has agreed to put the pouch on the train and the railroad people to put it off at Sunshine as soon as they get the official notification that there is a post office there. The Postmaster at Winslow is doing his best to get this straighten out at Washington. The post office at Canyon Diablo was called off for the technical reason that the Trader there carries mail to Leupp so may not act as Postmaster anywhere."

Navajo Blankets
Wool and Pelts
General Merchandise

H. W. SMITH, Mgr.

VOLZ TRADING COMPANY
INDIAN TRADERS

Canyon Diablo, Arizona

December 2nd 1920

Mr Holland Supt Crater Mine

Dear Sir and Friend
               Mr Fays Party are representive business men of Flag-taff who are well known to me   ?

  It is my honest beleif that they have a wonderful propsition and I am personally getting in for all I possible can handle

I heard some excellant news yesterday about this propsition from very reliable sourse and I feel sure that you would not make any mistake if you came in

Your realize Mr Holland that I would not recomm end this if I was the least bit dubious

        Yours very truly

I am very busy this morning else I would come over with them

Things are moving slowly forward, and they will be getting their mail without having to send someone or have someone come out to them over the terrible roads from Winslow. They are at Sunshine on a very regular basis. Their materials especially gasoline and drilling equipment but also food come to that station. And it is just 7 miles from the crater. The plans of the Volz Trading Company have been reversed, and they will not have a post office, but they can use the new Crater, Arizona P.O. at Sunshine. It is very close to them just several miles down the railroad tracks which usually have a service road next to the rails. It may not be what they wanted, but it is a great help to them as well. They were even farther from Winslow to retrieve their mail.

On February 15th Holland makes the following report regarding the post office. "The Post Office is now duly established at Crater (Sunshine Station) and the mail is sent by rail in a pouch from Winslow on Mondays, Wednesdays, and Fridays. Only occasional trips to Winslow will now be needed, and telegrams had best be sent to Canyon Diablo."

Holland makes no further mention of the post office in his weekly reports. It is amazing, though. The British superintendent of Crater Mining Company has navigated the waters of the United States federal government in 1921 and has gotten a successful outcome.

The Money

Six sons and I want them to all go to Princeton as I did. The Commonwealth passed its peak. They will still get more silver and gold, but my income was declining every year. I am glad we sold out in 1910. The claims in California have never been that rich. The iron in Mexico is still in doubt. I need to find the meteorite at the crater. Ten million tons in already metallic form. And with iridium, platinum and diamonds maybe it could bring $150 per ton. I could afford to gather a host of shareholders, split the profits and still achieve success beyond my dreams. But I can not spend much more of my money there. I have already spent over $100,000 on the hole, and we have still not found the buried star.

I heard from Jennings yesterday by wire. They turned me down finally last year after all that discussion. But they are interested again. I can offer them the same deal. A 99-year lease, they pay exploration cost up to $75,000, recover their investment with interest after we find the iron. I would get 25% after the discovery they would spend up to $600,000 more to start mining it out, and Standard Iron would get 50% after they are paid back. We all get rich. And I will deal with just the one group of businessmen. Assessing shareholders each month for more money and then worrying about their objections was very painful. I will organize any future companies differently. Maybe U S Smelting can find the meteorite. With ten holes and $75,000 to use in the attempt, I think we have a great chance.

It is under the arch of the south rim. I know it is. Otherwise, why would that part of the rim be so much higher and the wall there so sheer. The great eruption of the silica on the south slope was squeezed out by something. It has to have been the great asteroid. I'll have them do a new magnetic study to locate the best spots to drill. If the new study confirms the south rim on the MN line like the old survey, then that is where I must convince them to try first. 10 million tons but it is probably at least a thousand feet down from the top of the rim. But once we find it we can drive a tunnel to get it out. And tunneling is not that expensive.

Though Daniel Moreau Barringer had hopes of riches being found at Meteor Crater, it turned quickly into a huge financial drain. Within just several years of filing the claims, he had spent nearly $100,000 at the crater.

Everyone supplying the crater with materials or services was doing much better. The First World War had seen iron production skyrocket, but by 1920 it had returned to near prewar production levels. The price of iron ore varied but was generally in the area of $7 per ton. The price of pig iron was $28-$33 per ton. Heavy scrap iron for melting, what we would call recycled metal, was $15-$17 per ton. Barringer believed that the asteroid would be found in a largely metallic state ready for refining. The 7% nickel in the meteorite would raise the value of the meteoric material. Refined nickel metal in 1920 sold for forty-two cents per pound. Seven percent of a ton is 140 pounds. The value therefore of a solid ton of Canyon Diablo meteorite in 1920 might sound like it would be a mix of the iron value and the nickel value and the other even more valuable metals iridium and platinum. But, that would not be the case. The nickel, iridium, and platinum needed to be refined and purified from the solid metal of the asteroid once it was recovered. Furnaces were coal and coke-fired, and those fuels ran as much as $14 per ton. Also, processes had to be created to separate the metals. Johnson Matthey and Company world famous refiners of precious metals turned Barringer down several years after the drilling on the south rim. They did not know how to extract the platinoid metals from the iron at a profit and feared that the metal was in a single huge mass despite the lack of magnetism at the site. This was certainly one time when Barringer's ideas about the asteroid being a solid mass of metal turned around to bite him. The value of the iridium, platinum, nickel, and diamonds on paper was high enough to have Barringer thinking about $200 a ton. But, the reality was far different even if it was solid metal.

There is little doubt that there were solid meteorite fragments discovered during Barringer explorations at the crater. On several occasions, his drills in the crater floor were stopped by objects that could not be drilled. The objects were blasted with dynamite and did not shatter. Drill strings weighing nearly a ton dropped on the objects did not move them or break them but did ring with a distinctly metallic sound. Samples of the metallic material brought up from these holes showed the obstructions were metallic meteorite fragments. But, when a drill rig was moved only three feet the object was not hit again by the bit. Barringer's dream of a huge mass was never found, and on the south rim, the drill revealed mostly weathered meteoric material. At two spots during the drilling, they seem to have hit meteorite fragments. The drilling log of C. W. Plumb shows that he struggled for days to get through just a couple feet in a material so hard the bit was quickly dulled. There were strong indications of nickel in the chemical tests he did. But after just three or four feet he was again in the sandstone, with no nickel or iron.

If the iron was a bust, the silica of the crater was only slightly more successful as a commercial product. Barringer had Holland send samples of the silica to several companies and scientists during just his one year at the crater. For the most part, these were small parcels send by post. But Holland did send a ½ ton of the silica to the McKnight Firebrick Company, of Porterville, California. A history of that company shows that they did not need the silica. Their plant had been located deliberately near all the natural resources needed to produce their bricks. But, as often happens businesses will accept samples and do some experimenting to see if their product can be improved. Within a year McKnight Firebrick Company had merged with another brick manufacturer and closed the Porterville plant. Later after D.M. Barringer's death, his son would arrange a lease of the south slopes. Meteor Silica Corporation sold the silica for $3.25 per ton F.O.B. Sunshine Station. That operation would last for about three years. By the early 1950s, the Barringer family had control of the crater again. The deposit of silica on the south slope is estimated to be 1.5 million tons. By the end of the operation, Meteor Silica Corporation had shipped only several hundred tons.

Barringer had asked Mr. Holland to obtain prices from the freight department of the railroad. He wanted quotes for the shipping of silica from the crater to both the east coast and the west coast as well as the price for sending silica sand through the Panama Canal. On October 27, 1920, Holland gave Barringer the first of the price quotes he had received. The cost to send silica sand from Sunshine Station to West Chester, Pennsylvania was $1.10 per hundred pounds, and the minimum load was 80,000 pounds. If they sent only the minimum which was 40 tons, it would cost $880 just for shipping. It would have cost less to send to the west coast which may explain Meteor Silica Corporation sending what they did mine to Los Angeles.

The list of attempts to sell anything found at the crater is long. It includes using the sand for optical glass, as a cleansing powder, and more brick and tile manufacturing. But until the crater became the natural wonder visited by thousands of sightseers each year it was never commercially profitable. By the time the drill hole L. F. S. Holland began was finished by superintendent Plumb, Crater Mining Company had spent over $175,000 in about two years. Later explorations during the years 1928 -1932 would spend money at an even more accelerated rate.

Many of the suppliers did make money. The Republic Supply Company sold two drill rigs for a total of $30,000. The contractors though reportedly terrible workers and scoundrels were paid more than $10,000 for building the camp structures on the south slope. The wholesale grocer in Denver made money on the food. Standard Oil made money selling gasoline and oil to Crater Mining Co. The railroad got freight business continually.

Some other people did not make much for their services to Crater Mining Company. The delivery man who brought telegrams and other items from Winslow was paid a few dollars to brave the terrible roads. The pick and shovel men working on the drill site preparation and the pipeline made very little. Their hourly or weekly wages are not mentioned, but an estimate can be made. By taking the total expense, the length of time the work took to complete, the average number of men working and doing a little math, it works out to 40 cents an hour. The average wage for farm workers in 1920 was 60 cents an hour, but some workers in America were getting just 20 cents per hour. When Holland writes in his weekly report that some of the men have returned several times because they have likely spent the few dollars they were paid the week before. He really means just a few dollars. The men worked seven days a week and probably at least 55 hours a week which was normal for the period. They may have gotten only 10-15 dollars after the $1.25 per day board charge was taken. They would quit on payday go to town and spend the money and be back to work in a few days. Laborers were so hard to find that Holland took them back repeatedly. Some of the highest paid workers in America in 1920 were industrial workers. These factory workers made an average of $1407 per year or approximately fifty cents per hour if they worked the same seven days of eight hours.

Mr. Fay the surveyor was a regular contractor. He had produced a contour map of the crater. He later did a magnetic survey to determine how variations in the local magnetic field deviated from the Earth's magnetic field. He produced drawings and maps and was paid $1300 for just one part of his work. He was then called out later to do an even more extensive magnetic survey, and Holland says while it is still going on that he had been paid another $1000. This second survey took weeks for him to complete, so he received even more by the end. During the last days of February and the first couple days of March 1921, Mr. Fay was again at the crater surveying the location for the tunnel to reach the lost drill tools. More money spent for that. Mr. Fay and the suppliers of equipment and food were the ones who made money at Meteor Crater during the explorations of Crater Mining Company.

Drillers were paid $10 per day and not charged for board as an incentive to stay. Board was usually charged by the employer, and at the time of this work at Meteor Crater the amount was $1.25 per day in most of Arizona and $1.50 in California. However, the oil drillers in Holbrook, Arizona were not being charged board. Thus Holland felt compelled to follow suit matching his neighboring town down the road.

One of the most valuable workers on a drill rig was the tool dresser. These men worked at a forge and reshaped and sharpened the bits. They were paid at Meteor Crater $7.50 per day; a little less than the California rate of $8 but they were also given board. After all the problems Holland had with keeping workers, he could not afford to have the drilling crew leave over a few dollars. But, he was also motivated by his need to economize since the operation was nearly out of money before the drilling even began. Typically the drilling foreman would not work a shift on the drill rig and would get $500 a month. Holland got Mr. Wommack to agree to work one eight-hour shift as a driller and to take $15 per day and board. This was a bad deal for Mr. Wommack, but he did it. Even if the board charge had been taken out, he would have been a few dollars ahead at $500 per month and would not have labored on the rig. This agreement saved Crater Mining Company the cost of one driller and had a foreman working rather than just supervising. Holland pulled off quite a deal for his bosses with this arrangement.

Mr. Holland was paid $325 per month as superintendent. Considering his long list of responsibilities, this was a bargain for his employers and a low hourly wage to him. He was paid for eleven full months and a broken month of May 1920. His pay for the full months totaled $3,575. With the portion of his broken month and some pay for travel expenses, he made very close to $4,000 for his time at Meteor Crater. This was a lot of money for him. He was another of the men that did make money. On the census a few years later his income would be shown as $1300 for the year a much more average amount for the time.

The local rancher, Mr. Hart, was paid for some of the items he provided to Crater Mining Company while other items were loaned or given. He sold the company furniture for $172.50. There must have been at least one occasion when the supplies they had ordered did not arrive on time. For Hart sold the crater four sacks of beans for $29.40 and three barrels of gasoline for $54.00. That gasoline price would be right in line with what they were paying Standard Oil, so Mr. Hart did not mark the price up.

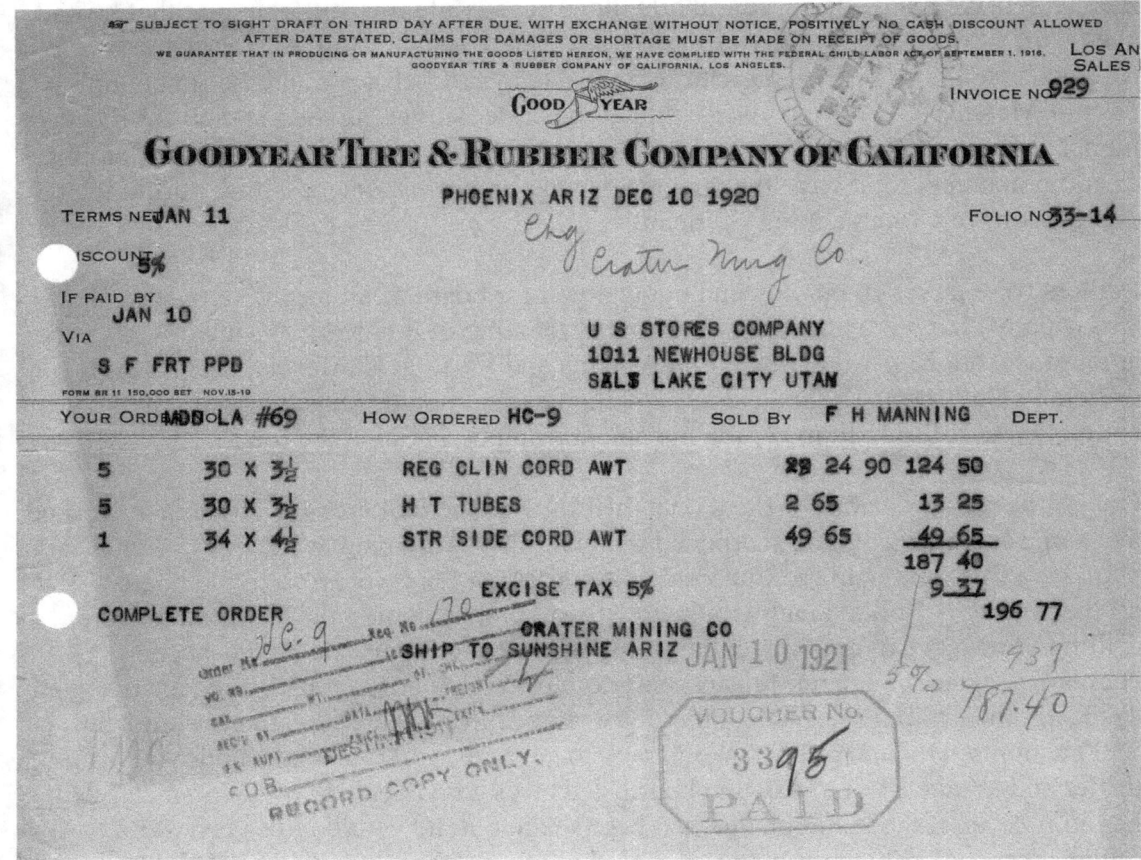

The Ford truck that Holland was using was bought for $500 from the Hart Cattle Company. New tires were purchased for the Overland car. A set of five Goodyear 30 x 3 ½ REG CLIN CORD tires were $24.90 each for a total of $124.50, but they did need tubes which cost an additional $2.65 each. Goodyear also supplied the respirators for the miners digging the tunnel. Crater Mining ordered six white respirators for $4.28 with a 25% discount bringing the cost to $3.21. It cost just 30 cents to mail the respirators to the crater.

United States Stores Company Inc. from their office in Salt Lake City was the purchasing company for many of the things used at the crater. Like Crater Mining Company, United States Stores Company was a subsidiary of United States Smelting Refining and Mining Company. It was through purchasing agents with requisitions that Holland had to order the tires, respirators and such things as his microscope, the Jones

Samplers, the typewriter and much of the bulk food such as cans of lard and sacks of flour and sugar.

Holland had always intended in his correspondence to gather up laboratory equipment from his field office in Los Angeles to use at the crater for nickel testing of the drill concentrates. But, he never made it to Los Angeles to get those supplies. Two of the several reagents were sent to the crater by Barringer. Holland used the bottles and containers he could find or save for the chemistry. However, a microscope was purchased from Bausch and Lomb. It was a "chemical microscope (new style) for use on opaque mineral work" and had a vertical illuminator bought with it. The total for the equipment was $153.25 with a 15% discount making the final cost $130.26. It was also obtained through U S Stores. Most of the bills to the crater offered a 2-5% discount for on-time payment by the 10th of the following month. Such discounts for timely payment were common in many industries in America through the next several decades. By the 1980's the business practice of having credit line accounts with individual suppliers was disappearing. Companies no longer had their own internal credit departments which approved credit lines and determined credit limits. Credit card companies emerged and monthly interest charges and time payment plans replaced the reward discounts for paying on time.

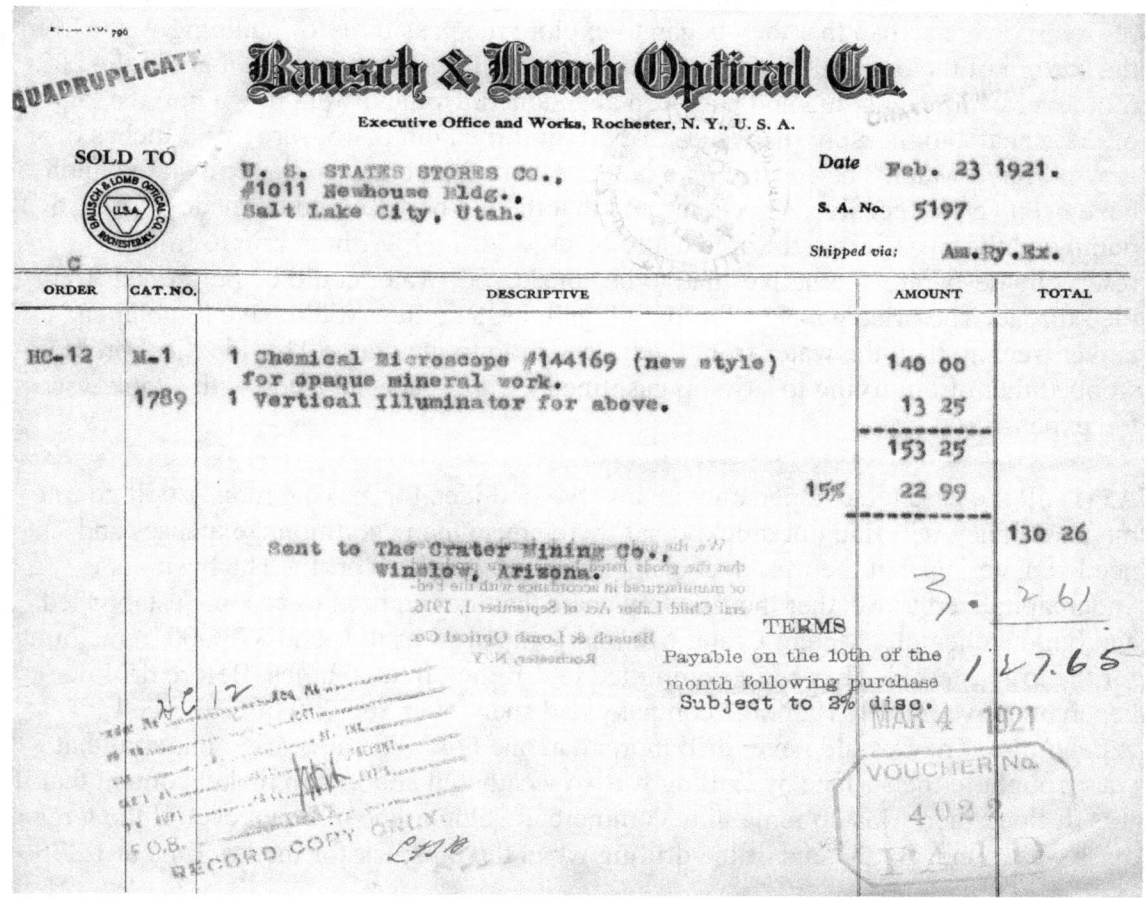

Though it has been nine decades since the supplies were ordered and sent to the crater, it is interesting the brand names that survive until today. Goodyear and Bausch and Lomb as mentioned but also SunMaid; they ate seeded raisins at the crater. Log Cabin shortening was used but no mention of their syrup. The sugar of the time came from beets. It was sold in sacks at $9.43 a sack. The Norton Company the world's largest supplier of abrasives still today received a sample shipment of the silica sand from Meteor Crater. It was sent by Mr. Holland to the large furnace facility that The Norton Company had at Niagara Falls.

Anything that had to be transported was expensive. Rock, for instance, had to be hauled for the foundations of the pump at the dam site and the mounting for the gasoline engine at the drill site. The gasoline engine foundation required ten cubic yards of rock and cost $300 for the rock to be laid down at the site. This material came from the Little Colorado River from a location suggested by D. M. Barringer. They chose to use local boulders for their later needs at a far lower cost.

During the last few weeks of Holland's stay at the crater, there were increasing concerns about the supply of water for the drilling. The drill required a constant amount of water down at the bottom of the hole. Compared to the amount of water that was required by the camp and the men, much more was needed for the rig. The conditions in the reservoirs were so bad that they began to explore the possibility of pumping water from the bottom of the original shafts in the crater floor. These had flooded at about the 200-foot level while being dug and had been abandoned. Holland went down into the larger of these shafts to measure the water level. As it turned out there were just 2 inches of very polluted water. The cost to run a pipeline out of the crater to the drill site would have been very expensive. They expected that the old boiler could be made to drive a pump and that wood from the area could be used for fuel. Without rain to fill the reservoirs desperate alternatives had to be considered. Water could be purchased from the railroad. The price was $35 for 10,000 gallons. But they would have to obtain a tanker truck to haul the water from the train station to the crater. This identical problem stopped them from trying to save on gasoline as well. A special truck with a tank was too expensive.

As is still the case today a person with a very good idea for making money will go into business. They will find out quickly that there are so many additional expenses and needed materials that their product cannot be produced for a profit. The business is undercapitalized to weather the first few years that are required to become established and build a clientele. Meteor Crater Mining Company was to spend $75,000 in the hunt for the iron asteroid. They were committed to drilling up to 10 holes. Before drilling began on November 1, 1920, the company had spent over $60,000 of the agreed expenditure. They would never drill more than one hole. The meteoritic material that was brought to the surface by drilling was so weathered and low in nickel content that it was decided there was no remaining commercial value in the material even if the large mass was found. By the end of the drilling when the bit stuck for the last time at 1,376 feet the parent company United States Smelting Refining and Mining Company had spent between $175,000 and $200,000 drilling the one hole. They would recoup some of their expenditure by selling off the surplus unused and serviceable used equipment. But that amount would be small compared to what had been spent.

Barringer's Footprints

D. M. Barringer is a character that cannot be separated from any study of Meteor Crater. For anyone obsessed with meteorites and Meteor Crater Barringer is an almost revered historical figure. When I opened the first of the file folders and saw the Daniel Moreau Barringer letterhead and the typed pages that he had personally dictated nearly a century ago I was thrilled. Then as I worked my way through the documents, I found many more pages on his letterhead, handwritten letters, notes and even some sketches drawn by Barringer. This let me know that his footprints were all over these lost papers. He also left his tracks all over the work done by Holland in his year at Meteor Crater. I held the pages in my hand and knew that the paper was his, the ink was his, and that it was his hand that had written the words with his pen. That was certainly as exciting as finding a meteorite. It is not a matter of hero worship. Barringer is not a personal hero though he is a fascinating historical character. There are many aspects of the man that this author dislikes. But, as a historical writer holding actual documents is as good as research gets. And Barringer was a prominent personality in the Meteor Crater story.

One of the first Barringer documents in Mr. Holland's files was a handwritten note penned on Wednesday, June 30th, 1920 at 8:30 am. Holland has been on the job at Meteor Crater for about six weeks. Barringer and his son Reau have been at the crater for five days when this neatly folded note is left for Holland and marked personal. Barringer would occupy Holland for the whole day, and the note was Barringer's schedule and a list of the things he needed Holland to do.
Holland has received at least two telegrams and likely three, but one appears missing about their schedule and arrival. On June 17, 1920, a Thursday, Barringer sent a telegram informing Holland that they were "most unfortunately obliged postpone departure until Monday which should bring us to Winslow on number one morning twenty fifth." This postponement telegram suggests that there might have been an earlier telegram which is not in Holland's files in which he is originally told of the Barringer visit. On June 19 the Saturday before his scheduled departure from Philadelphia he sends Holland a follow-up confirmation telegram that he "will arrive Winslow with my son on number twenty-one next Friday morning please arrange to get me out to the crater after breakfast" There are no punctuation marks in many of the telegrams. This one ends with a name "O M Barringery" a strong indication that it was a handwritten note which Barringer gave to the telegraph operator. Barringer used a final flourish with the pen which always swept down and to the left below his last name. From having seen his signature now many times, it is easy to see why the operator read it ending with a "y" and his very round capital "D" with a loop at the top right is easily read as an "O" also. The numbers mentioned in the two telegrams are train numbers.

It is likely that Mr. Holland had never met D.M. Barringer prior to his arrival on June 25th. Holland might have been a little in awe of the man. It was his crater that he was working at and it is a spectacular location. And Barringer had a big reputation in mining by then, something Holland likely knew. Barringer had been a principal of the Commonwealth Mine at Pearce, Arizona one of the largest discoveries of silver and the mine had produced significant amounts of gold as well. Holland had been a mining engineer for fifteen to twenty years by this time and worked in many areas of the western United States.

After this visit, there are a series of letters and telegrams from Barringer to Holland. It seems that some personal items have gone missing. Barringer is concerned with an umbrella and a cane that are important to him which he believes may have been lost at the Harvey House in Winslow and wishes Holland to investigate though he has communicated with the hotel himself.

Only days after Barringer and his son leave Meteor Crater Dr. and Mrs. Magie arrive for a visit. In a letter dated July 17, 1920, Barringer notes the receipt of Holland's letter of July 7 and says that he has heard from the Magies and they were thrilled by their visit and reports they "were very loud in their praises of your kindness to them. They certainly greatly enjoyed their visit to the crater." This must have been significant to Barringer for Professor Magie was one of his most influential advisors and supporters in the scientific community.

Later in this letter of July 17 Barringer reports that the missing cane and umbrella have turned up. "Reau placed them in the bottom of his long steamer trunk and forgot all about them. So we had them with us all the while and did not know it. I have written to the Harvey House expressing regret for the trouble which I caused them in making such a thorough search. I am also very sorry to have given you so much trouble with regard to the matter."

A little business is discussed, and then Barringer closes this letter with regards to Mr. Holland and asks that he be remembered to Mr. Philps and Mr. Fay and Mr. Smith. Mr. Fay is the surveyor who does all the magnetic studies and the site survey for the tunnel later. Mr. Smith is the manager of the Volz Trading Post at Canyon Diablo. Mr. Philps is an unknown character as far as Holland's papers are concerned.

Below the signature of this letter, Barringer penned this single line. "Arrived home yesterday pretty well used up, particularly my feet."

It is clear that Barringer and his son left the crater in the first days of July making the crater portion of their visit West rather short. On July 4th Barringer sends a note on Casa Loma Hotel stationary to Holland in which he first discusses in detail the travel plans of Professor and Mrs. Magie. He asks Holland to bring Mr. Fay back to the crater so that Dr. Magie can see the magnetic measurements being done for himself and get the best possible understanding of the deviations seen in the local magnetic declination. Barringer assumes that Mr. Fay is done with the survey. Professor Magie does get to observe the measurements and also makes some experiments himself while at the crater. Barringer also brings up the lost umbrella and cane for the first time and gives Holland an address in Tucson should they be found quickly otherwise he gives him the office address in Philadelphia. The Casa Loma Hotel was one of the quality hotels in Arizona at that time. It had suffered a devastating fire which burned it to the ground. It was rebuilt in 1899 and later remodeled into a southwestern style building in the mid-1920s. Barringer had brothers in law in the Phoenix area, and they had been involved with Meteor Crater since the beginnings in 1903. The original four claims were obtained in the name of the three Bennett brothers and Samuel Holsinger. Barringer was fearful that having his name on the claims might lead to someone "jumping their claims" since he was well known in Arizona mining circles. Once the claims were filed and final, little is heard about the Bennetts other than an occasion correspondence by one of them with Holland. However, they remained stockholders. Barringer would usually visit Phoenix when he came out west.

Almost immediately after finishing the first magnetic survey of the crater Barringer and the men in Boston have Mr. Fay begin another much larger magnetic study of the crater. During the month of August 1920, Barringer is frequently in communication with Holland. Barringer is also in discussions and meeting with the United States Smelting Refining and Mining Company officers. Elihu Thomson and Professor Magie will be kept in the loop receiving copies of much of the correspondence as does Mr. Holland.

On August 11, 1920, Elihu Thomson wrote a lengthy letter to Mr. Barringer in response to Barringer's letter about Mr. Fay's observations of the magnetic deviations seen at the crater. Thomson begins with a statement that he agrees with Professor Magie about the readings found at several of the stations on the crater rim. The reading at station 8 for example "is peculiar." The survey results seem to support their idea that the main mass of the asteroid came to rest under the south slope just west of the M-N Line. Thomson makes two pencil drawings of how he feels a third major deflection detected might represent the buried mass. This was the large deflection seen at station 24. The first drawing shows a mass that is lying inclined at an angle, with a tail that is closer to the surface and a larger portion buried deeper. His second sketch is the other possibility, where a mass of iron is low lying and which has spread or swept as it skidded to rest somewhat to the west. He marks on his second sketch where he thinks readings should be taken at five additional stations to refine the results. Thomson writes to Barringer that if the reading at these extra stations should show "a slight convergence – that convergence would almost surely point to the center of the mass – or at least show the general location..."

Unfortunately, this magnetic survey will never be detailed enough to satisfy the men that receive the data. They will, in fact, order another survey. The scope of this second survey is too large for one man and a helper to do to the degree requested. The locations where anomalies are seen and where locally refined studies should be done are sadly not performed during this survey either. The three best magnetic deviations found remained as single points of odd declination. The equipment was poor by today's standards, but some buried iron objects may have actually been located. Despite many letters like this from Thomson to Barringer and other letters from Barringer and Magie to USSR&M Co., Holland and Mr. Fay never get the careful local readings of these hot spots to detail their positions. USSR&M Co. will move forward with the first drill site on the top of the arch of the south side on the intuition that the spot is uplifted higher because the asteroid or the rock pushed in front of the asteroid has displaced that area upward.

Elihu Thomson
22 Monument Avenue
Swampscott, Mass.

August 11, 1920

Dear Mr. Barringer:

I am just now in receipt of your letter with Mr. Fay's chart of compass deflections. This, with Prof. Magie's letter I am returning as requested. I agree with Magie, in general. The reading at 8 is peculiar as he says, and may be due to a detached mass, local in nature. The other readings are consistent with a location to the west of line M-N about where I have faintly marked the map in pencil. The large deflection at 24 may mean one of two things; i.e. The iron may be higher up there (P.S. No there of course but where the deflection originates), a sort of trailing mass, or that the low lying mass is spread or swept as it would do if skidding somewhat to the west as we suppose it did on stopping - or to a combination of the two causes. I have tried to illustrate these conditions below. It would have helped much to have had another set of readings at stations such as a, b, c, d, e, outside and over the slope to the south. Also dip needle readings would have been excellent at such points; by which I mean read-

Elihu Thomson will finish this letter to Barringer with a lengthy discussion of the corrosion of iron and how the shale ball iron found at the crater likely formed. Barringer has been somewhat obsessed with this iron mineral product since the first days of his exploration of the crater. Many of the buried iron masses found in his early trenching were obvious meteorite fragments that had completed decomposed into the hard iron shale. Elihu Thomson offers an insightful look at how iron meteorites weather. He points out that the process begins with the fact that the iron is porous and that chlorine and other agents make their way into the iron where complex chemistry occurs ultimately resulting in the formation of the hard mineralized coating he calls "Ferric Hydrate (brown hematite)". The material is still called iron shale in the meteorite community today. Today meteorites that are experiencing corrosion are treated to remove the chlorine ions and break the chain of chemical reactions that Thomson has recognized and is presenting in this letter.

On August 19th just over a week after receiving the letter from Thomson, Barringer writes to C. F. Moore of USSR&M Co. about the magnetic survey and encloses another letter from Thomson. It is in this letter to Moore that D. M. Barringer asks that Mr. Holland and Mr. Fay be instructed to conduct the more detailed magnetic survey that he has been suggesting and which Elihu Thomson has said should be carefully studied after being made. Barringer says the survey can be completed quickly in a few days in his opinion. Barringer adds that the results can later be checked with a better dip needle instrument which is much more sensitive than the surveyor compass Mr. Fay currently has for use. Barringer makes it a point to say tthis more detailed magnetic survey will be of great assistance in locating the second drill hole. Barringer has held the opinion that the asteroid rests below the top of the south arch for 20 years. That site is already being prepared by Mr. Holland's crew of laborers. They expected that more than one hole and as many as the ten the contract allowed for might be drilled by USSR&M Co. It is a shame in one sense that the very detailed measurements of the several anomalies were not obtained. They could have been locations for modern researchers to check with tools available now. When the drilling work ends, many details like this are lost. The work at the crater moves on in other directions. It is interesting that Barringer is clear that he has no power to instruct Holland to do any work and that he must utilize proper channels and go through Holland's bosses in Boston. In this case, Mr. Moore is in agreement with Barringer that a detailed magnetic survey better than the simple one already completed is needed. Holland is quickly instructed to get Mr. Fay back out to the crater. Moore will also submit a detailed drawing of how he wishes the stations for the study to be laid out. He also requests that measurements of the magnetic declination be made along north, south, east and west lines out to a mile from the crater center to determine if a local deviation is seen or if any strange deviations continue to appear in the readings. Holland rehires Mr. Fay, and the survey begins. But the survey is such a huge project that it is weeks of work and even when completed is not as extensive as requested. This lack of depth in the survey puts Holland in some hot water with Moore who had given him exact instructions and a diagram. Barringer has gotten his survey, but Moore's recommendations were not the same as his. Barringer will still be disappointed in the completed survey along with all the others.

Just two day later on August 21, Barringer wrote Mr. Moore with additional information on the first magnetic survey. Barringer has not seen the maps which Mr. Fay was to produce, and Mr. Holland was to send. He makes a request that they are sent to him as soon as possible as he is eager to study them. But he states that he has received a message from Holland that all locations with deviation from the normal magnetic declination have been noted. That at any other location the declination is normal for the region. Barringer acknowledges that though slight, the deflections are real. Barringer again brings up the suggestion already made by others that a complete circuit of stations is established, and a survey of the entire crater and the surrounding be made. He uses slightly different words in many of the letters, but they state the same thing that he says the magnetic data will "be of great value to us if properly interpreted."

The letter that Barringer refers to in this message to Mr. Moore is Holland's response to the missing umbrella and cane note which was sent by Barringer. He enclosed a copy of Holland's letter and felt the need to explain the information about the umbrella and cane. Then at the end, Barringer takes up a brief discussion of what Holland has said about it being inappropriate for him to write an article for a magazine requested by Mr. T. A. Rickard. Holland has told Barringer that he thinks any article should be written by Barringer since he is the discoverer of the crater's true nature and not by someone who has come on the scene so late. Barringer writes on this to Mr. Moore with the following statement. "I think it will be pretty difficult for anyone to take away from me the credit of having discovered the true origin of the Crater, as my acknowledgment papers long antedate any papers which go to show this. As I have frequently said, the origin of the Crater is as much proven today as it will be when the meteoritic mass which made it is discovered." Still, the faithful believer that the huge mass of nickel-iron rests somewhere under the ground waiting for him to find.

One can only imagine the kind of response that Mr. Moore would have as he receives his mail day by day. For on the 27th of August Barringer sends off another letter to him. This is his response to the letter of the 23rd written by Moore. But this letter is an important one for us looking back. Barringer begins as he always does with a thank you to the person for the letter he has received and to which he is responding. Mr. Moore has sent instructions by letter to Mr. Holland as Barringer requested. Mr. Holland is to have Mr. Fay take measurements at 200-foot intervals around the crater rim. At any location where a deviation is read, there is to be a detailed survey of the location to define the extent and shape of the area of magnetic variation. Once again Barringer expresses his desire for Mr. Fay to have a dip needle rather than the compass. Barringer continues in the letter to describe how he would like to see the magnetic data recorded on the maps using red arrows showing the direction and amount of variation. Not yet satisfied by his statement a few lines earlier Barringer suggests it is "advisable to call up Professor Thomson or write to Professor Magie, or both and ask if they know where we can borrow or rent a delicate dip needle instrument, provided of course with a tripod, which I have no doubt Fay could operate as well as anyone else." Barringer continues with his story of using a crude dip needle device in the bottom of the shaft on the crater floor. He believes that though it was minuscule that he had seen a deflection in the instrument. It is clear that his desire to have a dip needle at the crater is to partially satisfy his questions about that unresolved study he did nearly fifteen years earlier.

General affairs of the crater are written about in the remainder of his letter. The dams and their need of repair and heightening and Barringer's agreement with Holland that something should be done to make the cabins at the crater warmer. Barringer writes that they had used heavy tar paper on the early buildings. He adds that the correct choice of a stove will ensure that the buildings will be warm enough during even the worst weather.

Before he can complete this letter of August 27th Barringer receives another letter from Mr. Moore dated August 26th. The mail service ninety years ago was doing a good job at getting letters from Philadelphia to New York and Boston in just one or two days. Barringer answers some questions about iron pipe left in the holes on the crater floor which came to Moore's attention. The pipe and other iron objects on the floor could negatively impact the planned magnetic survey. Barringer replies in a postscript to these issues raised by the new letter from Moore. Barringer says he was unaware of any iron pipe disposed of in the bottom of the central shaft. He admits to abandoning much pipe in the drill holes when the holes failed. He concedes that if this material will interfere with taking readings on the floor that the plans to make those readings should be discarded.

Mr. Holland writes to Mr. Barringer on September 12th to inquire about the disposition of the metal water towers out at the dams. Holland has been asked by the government for the use of at least one of the towers. Barringer responds in a letter dated September 20 that all the material at the crater are under the control of Crater Mining Company and he can do as he feels best but that reserving at least one of the towers for future use might be a good idea. He suggests that Holland defers making a decision on either tower.

Barringer takes the opportunity of this letter to remind Holland that he would like to hear anything that Holland may learn from the shipments of silica sand sent to the McKnight Fire Brick Company. He writes that he would like to know himself even though he assumes that Holland will convey this in his weekly reports to Boston. If it was not clear from other places it is now Mr. Barringer is being copied on all of Holland's weekly superintendent reports.

Barringer lets Holland know that he has received an acknowledgment from Mr. Sharples that the two samples of silica sand sent to him have arrived. These were additional samples that Holland had sent near the time he sent the 1/2 ton to the other party. Barringer remains in this letter convinced that the already pulverized and very pure silica "could have a good business worked up" to use the resource.

Barringer closes this letter to Holland by saying that Reau has just been in the office and joins him in kind regards to Holland. Typical of all Barringer's letters he is very cordial and warm in his remarks to Holland. Yet as history shows both these men and another son Brandon Barringer will later speak quite negatively about the superintendents of Crater Mining Company.

Holland gets a month of rest from D. M. Barringer until a letter arrives dated October 20, 1920. Barringer is concerned that he let Mr. Holland and Mr. Fay know that there is considerable iron pipe abandoned in many of the holes he had drilled into the crater floor. It may have been a weight on his mind. It had been discussed at length earlier with Mr. Moore of USSR&M Co. Barringer may have still been concerned that he might be blamed for wasting the surveyor's time with a study of the floor, and so decided to tell Holland directly that there is much discarded iron in the ground. Barringer made maps of his drilling programs available to USSR&M Co. Those maps were forwarded to Holland at the crater. The drilling logs from Barringer's early years of work on the floor were also sent to the crater. Barringer wants the men to know that should Mr. Fay's compass show marked deviations near any of the old drill holes "it is not at all improbable that it was due to the presence of the pipe in the hole."

Barringer moves on in the next paragraph of this letter to the drilling on the south rim. He has been told by USSR&M Co. that the rig is built and the drilling crew was sent for from California and that Holland expects to start drilling the following week. Barringer has waited for this moment for years and is clearly excited. He writes as restrained as possible the following. "I wish I could be there to see the start made and shall await with interest copies of your weekly reports regarding the progress which is made."

Just a week after this Barringer letter Holland receives a telegram about the peculiar visit of Dr. and Mrs. Campbell. Barringer was usually delighted to have scientists visit the crater. He believed that the crater was able to do a better job of convincing the skeptic of its impact origin than any verbal or written description. In this case, Dr. Campbell was not a skeptic. He had wanted to come to the crater in the spring with three very prominent other scientists but had been prevented by Barringer from doing so. The negotiations with USSR&M Co. had begun again, and Barringer feared any negative publicity that might again change their minds about entering into the lease. He did not know Campbell's opinion but did know that one of the guests Campbell had wanted to bring had just published an article which gave him pause. This telegram put Holland in an awkward position. He was to be a good host but was to impress upon the Campbells that they were being permitted to visit with the understanding that nothing would be written about the crater without receiving permission. Sidney Jennings,

Holland's boss, did not want publicity about the crater. Barringer instructed Holland to "be careful not tell him (Dr. Campbell) too much simply let him look around for himself." If Campbell had been frustrated about not being allowed to visit in the spring, he might have been even more offended by the cold shoulder that Barringer had instructed Holland to give he and his wife. But Holland was gracious and sent the driver to receive the Campbells from Winslow. He told them in a message before their arrival about the accommodations arranged for them. Holland apologized for not being able to spend much time with them because of difficulties getting the drilling operation up and running. The Campbells had an enjoyable visit. But Campbell who was in charge of the Lick Observatories of the University of California was actually a supporter of the impact origin theory. He seems to have never forgotten the rude treatment he had received from Barringer earlier. He never participates much on Barringer's side in the debates which raged about Meteor Crater's origin.

Holland receives a lengthy letter dated November 4, 1920, from Barringer. He is responding to the last two letters of Holland. Barringer had requested that Holland obtain quotes for shipping the silica sand to the east coast and the west coast and through the Panama Canal. Barringer had also asked Holand to send samples of the silica to a Mr. Sharples for experimentation. Mr. Sharples has determined that the cost of shipping the sand from Arizona to the east coast is prohibitive but has said that it is far superior to the sand available in the east. Barringer was dogged in his determination to sell something from the crater. If it was not yet the asteroid metal then why not the millions of tons of silica sand. Barringer goes on in the letter to say that the inventor of the method of making bricks and tiles from the silica sand had said the same thing about it. The silica sand is really just crushed sandstone. It is essentially the same as other sand except that it is as fine as flour. In some processes this would save the manufacturer the step of crushing and grinding down the rock. It is quite pure so the process of treating the raw material to cleanse it of foreign matter and other rocks would be eliminated also. Barringer uses the words "unduplicatable material" in his descriptions of the silica on several occasions. The cost of shipping it far exceeds the saving gained using it instead of a more local material found in the east or the west coast. There is also the problem of the iron it contains from the asteroid impact. The iron contamination had made the crater silica unsatisfactory for optical glass manufacturing during the First World War. Additionally, the government had at the time sufficient supplies of other silica.

After the discussion of the silica Barringer relates that he has received a copy of Holland's telegram sent to Mr. Jennings about starting the drilling on November 1st. He is clearly eager to see them get down to the asteroid. After nearly 20 years of believing that it rests below where they are drilling, he is writing with some excitement. He expresses here a hope which will become the last straw clung to by everyone in the future. The hope that the drilling will be better after they pass through the hard limestone and reach the gray and white saccharoidal sandstone. Barringer acknowledges in the following quote that it might not happen, though. "That is unless you have trouble because of the cracks and crevices which must be in it, as well as in the limestone, as a result of the uplifting of the arch."

After this cordial and congratulatory beginning of his letter, Barringer switches moods and raises the topic of the magnetic study. Here is what he wrote. "I was very disappointed in Mr. Fay's magnetic map. I thought this would be a map of only the southern third of the crater and that the stations would be on 200' centers so that we might draw certain curves of magnetic attraction in the manner indicated to us in a letter received from Professor Thomson. I feel that stations taken at random all over the crater, inside and outside, and even up against the stone museum will not be of as much value, as such a map as I have indicated, where the station would have been 200 feet apart at the intersection of north and south and east and west lines. Let us hope that such great physicists as Professors Magie and Thomson will work something out of it. All I can glean at present is that most of the 15's plus are about where we believed they ought to be." Barringer inserts a handwritten marginal note at this point which reads. "I have not as yet had the time to study it carefully" beyond this point Barringer's handwriting on the final four words of the marginal note is unreadable. His original typed letter continues the discussion of the readings at the various stations. "The others may be due to irregularly scattered but hidden masses of iron more or less near the surface. I have always thought we struck one of them in one of the drill holes near or rather in the corner of #3 shaft. Certainly, we struck an obstruction which the drill could not penetrate although we tried to blast it out of the way. It must have been small for the next drill hole three feet away did not encounter the obstruction."

There is a spot in the above Barringer quote that demonstrates his self-serving attitude and his very narrow vision. "I thought this would be a map of only the southern third of the crater and that the stations would be on 200' centers. . ." He has ruled out in his mind the possibility of the iron mass being anywhere other than the southern portion of the crater, and he wants the survey to detail the portion that is near where he thinks the mass resides. The men in Boston have a wider view and are seeking to discover the entire crater's magnetic characteristics. Barringer has had a closed mind for some time. It is a discouragement to him that the efforts of the magnetic survey are not centered around his wishes. He also ignores by his wording that Mr. Holland and Mr. Fay are working from a very specific plan. Barringer refers to the stations as "random" knowing all the while that Moore diagramed them for Holland to follow.
Barringer closes this letter with "I hope that Dr. and Mrs. Campbell enjoyed their visit but do not see how it was possible for them to get much of an idea of all the problems in a single day." He clearly knows that the reason they might have unanswered questions is his fault. He explicitly instructed Holland to "just let them look around on their own and not tell them much." Barringer signs the letter and then puts another note after regarding some photographs Holland has said he enclosed in the letter of October 28 which were missing. He then initials the letter after this final note. All these marginal notes and handwritten corrections and insertions along with the initials add some richness to the documents. It gives them an extra bit of excitement factor for a

historian. It is one thing for him to dictate the letter as he did, and another thing for him to have gone back and made all the hand additions to the pages in his almost unreadable handwriting. That makes them very special these decades later. But in this letter, he is not 100 percent honest with Holland and is dealing and complaining behind the back of the men in Boston.

Sidney Jennings sends a short letter to Holland dated Christmas Eve 1920. It is little more than just an acknowledgment that he is enclosing an extract from a letter he has received from Mr. Barringer two days earlier. Barringer is very interested in knowing what happens to the water level in the hole when the drill reaches the water table which he struck on the crater floor. Barringer has long held the belief that the water is trapped in what he calls an "ensealed zone" and that water will not be found at the same depth beyond the crater floor. The drill on the south rim is his opportunity to get the answer to that question. He would also get final closure for the reason the shafts failed because of quicksand encountered at roughly 200 feet. Jennings is likewise interested in the answer. Water soaked asteroid iron would have suffered considerable decomposition and as he already knew the iron shale was very depleted in nickel and of little value. Barringer has some supporting reasons for his belief that the water is trapped in the confines of the crater interior. The railroad had failed to find water with a drill just a few miles north. Holland is instructed to keep the men posted about the water in the drill hole. This would be an easy thing for the men to notice. During the drilling, they were required to put water down the hole to maintain a long column of water at the bottom. As they used the bailer to bring up the mud of pulverized rock the water would have to be replaced. If they hit the water table, they would from then on essentially have a well that would be full as long as they stayed in the aquifer. They would no longer add as much water from the surface. Such a change in the work process would be immediately seen by the drillers.

D. M. Barringer does not correspond with Mr. Holland again until January 28, 1921. At which time he is clearly aware of the difficulties that Holland and the drill crew are having. He receives copies of the reports which mention iron in the bailer and specifically the report of Christmas Day stating some small pieces of shale ball were brought up. Barringer asks Mr. Holland to send him some pieces of this shale ball material. Regarding the other "iron" that is mentioned he states, "I cannot help wondering whether this is also in the nature of shale ball iron oxide. Of course, no metallic meteoric iron would be brought up, and if metallic iron, it would probably be pieces of the lost tools. I am very anxious to subject these specimens to the dimethylglyoxime test. If they react for nickel they are certainly meteoric in origin. If they do not, they are probably portions of an oxidized iron pyrites nodule. . . If the iron oxide brought up by the bailer reacts for nickel it is not a little encouraging and it is not difficult to see how small meteorites would have been hurled into the face of the cliffs at the time of the impact, especially as there must have been many large openings in it as it was being formed and raised, into which individual meteorites from the upper portion of the impacting mass and somewhat separated from it could readily have found their way." What a long sentence. This man is clearly excited about his topic and is expecting to have all his questions of the last twenty years answered. Holland did the nickel test on the shale ball meteorite fragments during the drilling. The rocks were indeed very broken and creviced. A big problem for the drillers. It was not so encouraging as Barringer has made it out to seem concerning the shale ball meteorite pieces. The crevices in the rock were filled with a mixed material from above that had run down. Even near the bottom of the hole long after Holland departs, Mr. Plumb will still be hitting chunks of red sandstone, seashells, and bits of meteoric iron shale. Barringer's belief that these pieces of shale ball meteorite are remains of meteorites thrown into the rocks is probably wrong. More likely the bits of iron shale are part of a mixed material that fills the crevices. They do test positive for nickel and are without a doubt meteoritic in origin. As for the rest of the iron that is coming up in the bailer, it is from the lost tools and casing that are being gradually beaten up by the drill bit.

Barringer ends the body of this letter by telling Holland that "it is not improbable that I shall have to go west next month, and if I do I shall, of course, arrange to go to the Crater and pay you a visit." He adds in pen and ink after his signature "Moreau asks to be kindly remembered to you." Daniel Moreau Barringer Jr. usually used Reau for the nickname of his second son. It would have been a convenience to call him by something different as he was D. M. Barringer the third. Holland replies to Barringer's letter on February 3, 1921, as soon as he receives it. He writes that he is sending him "the few tiny pieces of shale ball from the 256 foot bailing inside of an envelope in a box containing a sample of the bailing itself. You may be able to find some more pieces with the aid of a magnifying glass."

Holland continues in this reply letter "I am also sending you in another box a sample of the 288 foot bailing where the sandstone commenced to show up with the lime. This is the material with considerable steel and iron from the tools referred to in my wire to Mr. Jennings. I got no reaction for nickel with the dimethylglyoxime, though I precipitated the iron from the acid solution with both the ammonia and with acetic acid. If you can have another test made on the material I shall be very glad. In the absence of suitable apparatus here, it is not convenient to make chemical tests. The first time I go to California, I will bring back some of my field laboratory outfit, but I have had no chance to go anywhere but the Crater and Winslow in many months. However, I think I am obliged to go to California soon to stir up the people who supply our tools and repairs." Holland goes on to discuss the problems of not receiving timely shipments of fishing tools and drill stem repairs. He is happy to report that they did finally receive one of the four broken stems back and that they were again drilling in the sandstone at the 307-foot level. He could not have known that just five more feet of drilling would be as deep as the hole would ever reach under his supervision.

By the third day of drilling on November 3, 1920, the hole was already crooked and in need of being redrilled several times to straighten it. Since then the crevices have given the drillers nothing but problems. The main challenge being to keep the hole straight so casing will go down and so drill strings will not break. On February 9, 1921, Sidney Jennings forwarded to Holland one of the most interesting documents in the entire collection. It is a small piece of paper with a hand drawing done in pencil on one side and brown pen ink writing on both sides.

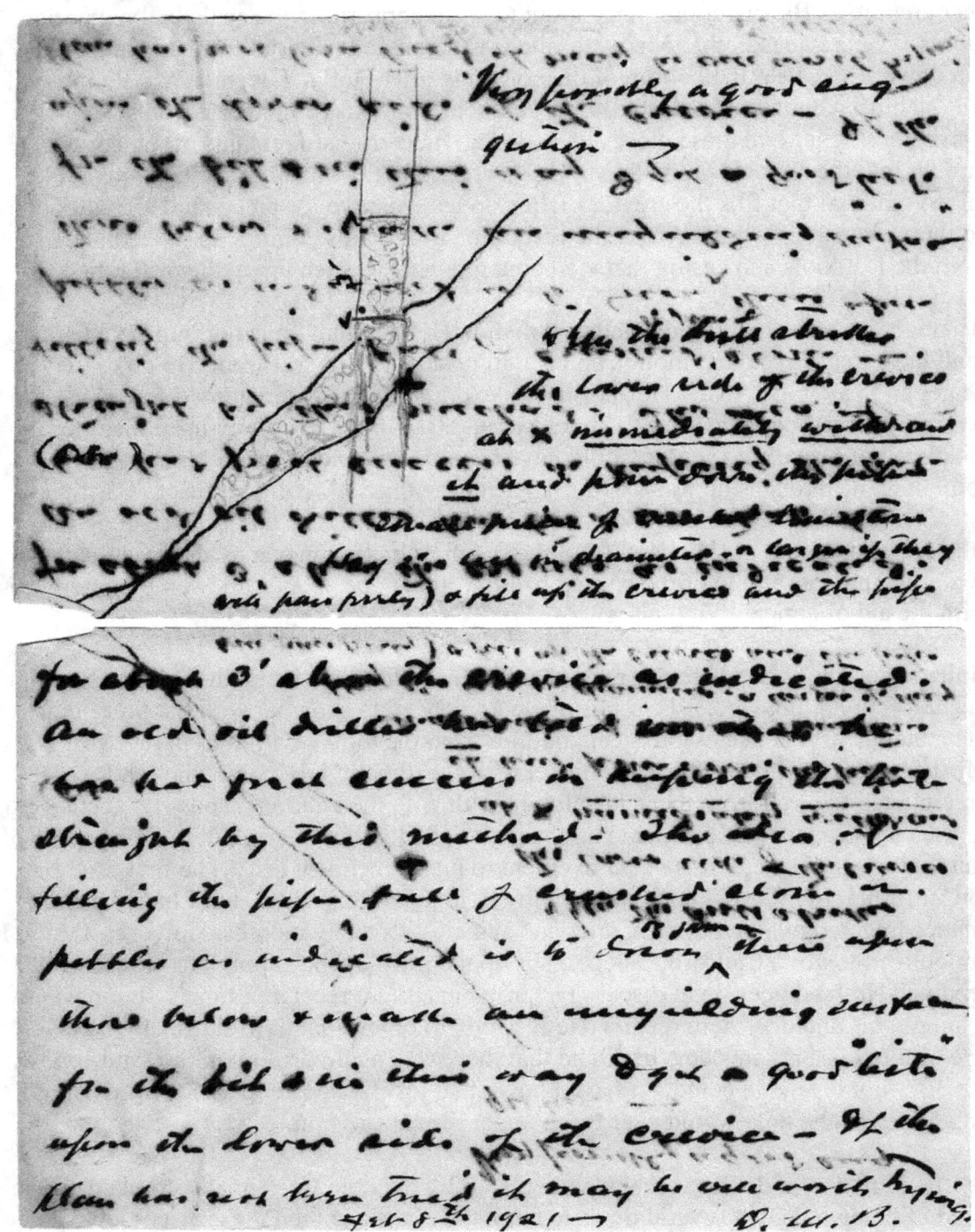

It is unfortunate that the writing is so small and the bleeding of the ink so bad that it is partially unreadable. But it was created by D. M. Barringer and intended to assist Mr. Holland in drilling the creviced rocks. The drawing is of a crevice with a drill hole crossing the crevice. The contact point of the bit on the bottom side of the steeply sloping crevice is marked with an X the drill hole is shown in the drawing filled with material. Of the portion that can be read, it is clear that Barringer is suggesting that the drill is lowered into contact with the steep crevice and then raised and lowered again and again. This is to be repeated until a shoulder can be cut into the side of the crevice. This should keep the bit from glancing off and bending down the crevice until it breaks. He is also suggesting that the hole be filled with crushed rock and pebbles. The end of his writing is somewhat more readable and states "create an unyielding surface for the bit and in this way to get a good 'bite' upon the lower side of the crevice – If this plan has not been tried it may well be worth trying. Feb 8th 1921 D.M.B."

Once again it was extraordinary to have an actual pencil drawing with handwritten descriptions and instructions made by D. M. Barringer himself to examine. But the real purpose of this letter from Sidney Jennings was to inform Mr. Holland that a "Mr. Keady desires two small sample lots of ten pounds each of the silica, be sent as follows. One to: Norton Company, Niagara Falls, New York marked attention of Mr. P. G. Savage. And the second one addressed to Dyke V. Keady, Melrose Highlands, Massachusetts." Barringer has been in conversation with Keady who has shown considerable interest in the utilization of the silica sand. Mr. Jennings is interested because there is still a chance that they will continue a lease of the Crater and would be in charge of supplying the material. Holland is asked to show Mr. Keady every courtesy and give him all the information he has on the silica if he should visit the crater.

By the time that Mr. Holland receives the letter dated March 29, 1921, from D. M. Barringer, the drill hole has been declared lost at 312 feet. Barringer has not gotten his answers to any questions. They are over a thousand feet short of reaching where the asteroid is thought to rest. Holland is digging the tunnel to recover the lost tools and hopefully clear the hole so drilling can resume. Barringer is again offering his advice to Holland.

While much could be said of D. M. Barringer being solely motivated by financial gain in his actions regarding the crater, there is another side to the man. He has on several occasions shown himself to be a responsible and concerned human being. He has had a conversation with Jennings about the tunnel digging and the use of respirators by the workers digging it. He has taken it upon himself to contact the Bureau of Mines in Washington D. C. regarding a respirator invented by them. He encloses a copy of the letter he received back from the Bureau of Mines in his next letter to Holland. The names and addresses of three suppliers of the respirators are included in this forwarded letter. Barringer is much concerned for the workers at the crater. He relates to Mr. Holland some of his experience with the rock dust from years earlier in the following quote. "This much is certain – that those who work in an atmosphere filled with this fine, sharply angular silica dust should wear respirators of some sort. I remember that, years ago, when we were digging out the shale balls around cut No. 10, some of the men were so much affected by it that they began to spit blood, and it is the part of prudence, therefore, to furnish them with every possible protection against inhalation of this fine dust with sharp cutting edges."

Barringer changes topics and asks about the metamorphosed vesicular sandstone. Holland has described it using Barringer own name "Variety B." Barringer wants to know if any of it is stained with iron or nickel. He states that he would be glad to receive a little piece of it. He points Holland to the rock samples in the stone museum on the north rim of the crater and tells him that if he finds such stains they will show a reaction to nickel when subjected to the dimethylglyoxime test.

The digging of the tunnel was a severe delay in the drilling program, and Barringer is still waiting to get answers about what is below the south rim. Yet he sees the tunnel as an opportunity to get some new information about the crater that has never been obtainable before. None of his explorations into the talus of the crater or the crater wall were ever as deep as this tunnel will be. Barringer with some keen insight realizes that when the diggers reach the layers of sandstone behind the talus that valuable information about the rock's condition will be available. He wants to know "how much the stratum is shattered and whether there is much, if any, metamorphosed sandstone, etc." It is reported elsewhere in Holland's papers that two intact shale ball meteorites were found just at the point the talus is passed and the crater rock face is reached. Barringer was thrilled to hear this.

This was the last letter that D. M. Barringer would send to Mr. Holland. Considering how critical he and two of his sons will be of superintendents Holland and Plumb later it may be of interest to see how he closes this letter.

"Reau, who is home from Princeton for his Easter vacation, joins me in kindest regards and in the hope that the worst of our troubles are over. Yours very truly, Daniel Moreau Barringer"

Mr. Holland sent a final letter to Mr. Barringer on April 23, 1921, and it seems important to present it completely.

Dear Mr. Barringer: –
I am sending you a copy of the photograph referred to in my weekly report of April 11 which you may like to have. You may be able to observe that the tunnel mouth is only a little to the east of your favorite site for a derrick. We are continuing to get a lot of running sand which is very difficult to handle in driving the tunnel. I am sending you a piece of red and white conglomerate sandstone of which there have been several bunches.
It has been blowing tremendously today, and the old shaft house came down this

morning after weathering many such storms probably. I am glad it did not blow down while I was down the shaft yesterday investigating the water of which I found only two inches in the bottom, or I might have been in a very uncomfortable position. The boiler room annex and part of the end of the shaft house still stand.

I am sending you a digest of the log of our first hole. It reads like hell but was actually worse than that to me.

Please give my best regards to Reau. I have been having a visit from a very attractive Princeton graduate named Sinclair Armstrong whom Dean Magie did me the favor of giving a letter to me. He worked in the dusty tunnel as long as I thought good for him and appeared to like it. He is now riding with Hart's outfit until he goes to Wyoming. There is the possibility that I may have to make way here in a few days for someone who has been longer in the service of U.S.S.R.&M. Co. following the shutting down of several of their properties. In the process of sorting out the sheep and the goats, somebody has to be the goats and perhaps I am it. If this untoward event really happens,

"God be thy guide from camp to camp: God be thy shade from well to well; God grant beneath the desert stars thou hear the Prophet's camel bell."

    Very truly yours,
                    L.F.S. Holland

L. F. S. Holland's Own Words

Preserved along with the rest of the day to day paperwork from his time at Meteor Crater is Holland's manuscript entitled "Drilling for Meteorites." It is unknown if this article was ever printed. A search of the mining and engineer journals and other publications of the time has never revealed that it was. Holland kept two handwritten drafts, and two typed revised versions. The final version was divided into two parts, and he uses the phrase "as described in our last issue" in the first sentence of "Part Two." So it is an open question whether he had an offer to write a two-part article on the drilling at Meteor Crater and if he actually had it published. As will be seen later the manuscript has the feel of a lecture to a live audience more than an article written for print.

Holland begins "Drilling for Meteorites" as many other writers have including Barringer with a description of the local setting of the crater and its dimensions. To Holland, it was another great natural wonder in Northern Arizona right along with the Grand Canyon. He writes "it is a round hole three-quarters of a mile in diameter, and after partial filling from 500-600 feet deep. It is variously known as Meteor Crater, Coon Mountain and Coon Butte. Piled around the rim of the crater, over a hundred feet higher than the horizontal formations of the plain, are some two hundred million tons of rocks, many weighing hundreds of tons each, and millions of tons of white "rock flour." Altogether the debris represents more dirt than was dug in the Panama Canal." It is clear that Holland for all the difficulties he experienced in his year at Meteor Crater was captured by the wonder and magic of the place.

A glimpse of Holland's audience for this article shows in the second paragraph where he writes "The Crater is of more than ordinary interest to several present Calpet men who worked on the property. In 1920 the United States Smelting Refining & Mining Co., which was closely affiliated financially with the Ventura Consolidated Oil Fields. . ." Calpet was the brand name of California Petroleum Company which had purchased Ventura Consolidated Oil Fields Company, which included Ventura Refining Co. Calpet kept the product name Ventura Motor Oil. Calpet acquired Ventura Consolidated Oil Fields Company in 1926. The Texas Company later known as Texaco would buy Calpet in 1928. Calpet was famous for having beautiful and exotic super service gas stations. Their location on Wilshire Boulevard in Los Angeles had a two-day grand opening in the 1920s with many of Hollywood's famous actors and actresses in attendance. The female employees were dressed in movie style harem girl uniforms, and the male attendants pumped the gas in white shirts with bow ties. The station buildings were a Moorish Revival style architecture. Calpet published maps that folded out and also maps as stapled booklets. They were travel guides for day trips around California. It seems that Holland is writing to an oil industry audience. He will later list names of men from California who worked at the Crater. They were well known apparently to the "readers" or more likely the hearers of his presentation. He adds some personal notes about some of the men. This would be of no importance in a regular magazine publication. The men had been supplied to him by W. E. Coan of the Oak Ridge Oil Company a company acquired by Ventura Consolidated Oil Fields Company. It has the feel of being an article submitted to a company newsletter or a trade association newsletter more than an article for the general public's consumption. It has even more of the feel of a lecture to a live audience where slides are being projected on a screen and described by Holland.

In this manuscript, Holland finally names some of the men who came to work with him at Meteor Crater. During his year at the crater, Holland wrote about the foreman and the drillers and tooldressers and how much they were paid and even when some left for Christmas holiday. But, Holland does not use their names except to call the foreman, "Wammock." In this article, he shares with us the crew members names. Foreman Wammock's first name was Ike, his son who had some difficulties and was described as being of low intelligence by Holland was Howard Wammock. This article was written at least several years after Holland left the crater for he includes information about the end of the operation and Barringer's reports of the results achieved. In this article, Holland writes that Ike Wammock "has recently barely escaped with his life when oil well Oak Ridge Willard Number Eleven caught fire." Again these are details that would be meaningless to the general public.

In addition to Foreman Wammock and his son, we are told that the Meteor Crater crew included Les Currier, Bill Snow, the DePriest Brothers, Walter Linville, and a man just called Buckman. The audience Holland is writing to would seem to be aware of the full identities of the DePriest Brothers, and the first name of Buckman. These men all got a "rest" as Holland put it while the tunnel was dug to clear the lost tools from the hole. Holland confirms in this article that the cable tool drilling equipment was converted to a rotary rig after the hole was cleared. But, he falls short of telling us whether the "Ventura Boys" as he calls them were the men brought back when the drilling restarts.

This writer's first hike around Meteor Crater was as a teenager. We started where the short tour begins still today at the side door exit of the museum building. When my parents and I arrived at the south slopes, we found a dead steer apparently killed by a mountain lion. There were tracks in the sandy soil everywhere near the remains of the animal. Even at my young age and with all the excitement being at the crater I realized my parents were unsettled by this during the last half of the hike. Holland writes "Linville may be seen in the little group pictured with the wildcat they bagged at the Crater. Though the Crater project itself was about as wild a wildcat as any of us had ever seen, the live kind did not bother us much, but I hesitate to say how many rattlesnakes we killed daily for fear of being mistaken for Baron Munchausen's understudy. . . Another of our local specialties was the wind. The roofs of all of the camp buildings had to be held down with cables attached to "deadmen" buried deep in the ground. The cables over the "museum" roof were attached to big meteorites after part of the building had been blown away."

With this Holland concludes the first portion of his article. It has been all overview and stage setting with an introduction of the drillers. Holland moves onto the crater's history, the drill work, and the crater geology. He begins the body of his article with a couple of generalized statements that were probably not true even in his time. But, Holland was a mining engineer and not an expert on meteorites. First, he writes, "It happens that in and around the crater more iron meteorites have been found than in all the rest of the world put together. Even at that, they have been used, so far only as museum specimens and not for commercial metal. Those I have seen vary from 1800 pounds in weight down to the size of a pinhead."

The truth is that Meteor Crater even as the best-preserved impact crater on the Earth has produced only a small weight of recovered meteorite fragments. Most lists today put the total amount recovered around the crater at thirty tons of meteorites. Long before the time, Holland wrote this article the Cape York meteorites had been discovered. Ahnighito the largest one of these Greenland meteorites had been brought to New York City. The Cape York meteorites weighed as a group far more than all the fragments found at Meteor Crater. The vast strewnfield of Gibeon meteorites had been found in 1838 in Namibia. 13.9 short tons of those iron meteorites had been dumped in a single pile in the Public Garden in Windhoek, and many more had been shipped away. By around 1920 the Hoba iron meteorite was discovered in South Africa. At an estimated 60 tons it remains the largest meteorite known. As early as 1921 the possibility of extracting the 10 tons of nickel in the Hoba meteorite was being considered. Holland missed the mark with the first of these statements.

As to the range of sizes represented at Meteor Crater he was a bit closer to accurate. The largest Canyon Diablo ever recovered is the Holsinger specimen, and Mr. Holland was well familiar with it. He often saw it in the old stone museum on the north rim of Meteor Crater. It is possible it had never been properly weighted as of the writing of his article. It weighs 1409 lbs. (639 kgs). His statement of 1800 pounds for the largest crater meteorite was possibly the estimate being used at the time. Neither Barringer nor Holland had known about the most abundant meteoritic iron material at the crater. That being the tiny nickel iron spheroids that remain in the soil. There are trillions of them within two miles of the crater, and each is much smaller than the pinhead-size Holland mentions. The spheroids total thousands of tons of remaining iron asteroid material. While they are not primary meteorite material having condensed from vaporized asteroid, they might seem to restore some truth to his first statement about the crater's iron meteorites exceeding the total for the rest of the world. The spheroids were not correctly described until much later in the 1940-50s.

Holland gives a composition of the meteorites and a description of the material. "The iron meteorites have a consistent composition of about 8 percent Nickel, two tenths of an ounce of Platinum to the ton, small quantities of Iridium and other rare metals, the remainder being iron. As metal the material would be worth about fifty dollars a ton. It is harder than any armor plate, so that it would be impracticable to mine a single very large piece if such were to be discovered." Given the methods of the time Mr. Holland's pretty close to the right numbers for the composition of Canyon Diablo meteorites. The masterwork "The Handbook of Iron Meteorites" by Vagn Buchwald 1975 reports the following amounts for the elements: Nickel 7.1% Cobalt 0.46% Phosphorus about 1% Carbon about 1% Gallium 80 ppm, Germanium 320 ppm, and Iridium 1.9 ppm. Buchwald goes on to point out however that there is sufficient variation in the scores of Canyon Diablo meteorites analyzed to support a Nickel content range of 7.0% to 8.2% which would make the Holland 8% an accurate figure. Barringer obtained a 7% nickel and 10 grams of Platinum per ton analysis a hundred years ago, and the amounts appear to still test in that range. Holland's price of fifty dollars a ton is much more reasonable than the $150 or more per ton that Barringer was often repeating to prospective investors.

Harder than armor and impossible to mine as a single piece are the exact reasons that Johnson Matthey the London-based precious metals company declined to enter into a lease with Barringer. They did not know how to get a huge mass of solid metal out of the ground, and they did not know how to get the small amount of Platinum out of the iron alloy it was locked in.

Holland never struck any metallic meteorite with the drill bit, but he ground up and brought to the surface a little of the shaleball material. This is a weathered mineral product that the asteroid has turned into over the thousands of years it has been buried. Holland makes the correct report that the iron shale had somewhat less nickel, iron, and platinum than the metallic meteorites. It was later proved by the chemists at USSR&M Co. that the depletion of the metal was so severe that the material was essentially not recoverable at a profit. They turned back their lease to Barringer. Barringer did not take that analysis to heart. He seemed happy to have gotten control of the crater back from USSR&M Co. He then begins the most expensive of all the explorations for the asteroid. He has only the small amount of shaleball material found long after Holland left, in a thirty-foot zone near the bottom of the hole to suggest anything meteoritic was buried where they drilled. Holland had seen plenty of the iron shale. It was very abundant in the early years. Thousands of the small pieces were collected, and it was still seen all around the crater during his time there. Holland saved the material brought up in the bailer and when it was drained of water he separated any iron from the debris with a magnet. He would test these little bits of iron shale with the reagents he had on hand. The iron shale had always shown a nickel reaction for Barringer, and it also did so for Superintendent Plumb. Holland just never recovered more than a few tiny bits of iron shale in his bailings.

The presence of diamonds in some of the meteorites and iron shale from Meteor Crater had become a news item. In his article, he makes light of what we would call today a tabloid article about them. Holland referring to the meteorites and the shaleball material writes "Both forms contain very tiny diamonds which can be seen with a powerful microscope. However, the alleged photograph of one of these meteorites published in a typical Sunday Supplement, showing an embedded diamond proportionately about the size of a decanter stopper, is merely a yellow journal's method of conveying scientific "facts" to its readers - and incidentally insulting their intelligence. As one of the Ventura Boys wrote me, it is too bad we did not get this "information" while we were on the ground so that we might all be wearing diamonds."

The diamonds are as Holland writes very small. They are black in color and called carbonados. They are without exception found only in the heat altered and shocked meteorite fragments found on the crater rim and the shaleball meteorites buried in the slopes of the crater. The diamonds formed when the iron meteorites were heated by the energy of the impact and the graphite they contained converted to diamonds as tremendous shock waves moved through the asteroid fragments. The meteorites found farther from the crater are often pieces of the asteroid which tore off in flight through the atmosphere. Undoubtedly many are also pieces thrown out of the crater but which did not experience the tremendous heat and shock waves. Some added value might have been derived from the recovered asteroid, if the diamonds could be separated and collected. Industrial diamonds were finding wider use about the time of Holland's stay at the crater.

Holland moves on by telling the story of D. M. Barringer and all of his activities at the crater, from his first unsuccessful shafts through to his drill holes into the floor of the crater. Holland writes that there were twenty-seven holes drilled into the floor. Everywhere else records the number of holes as twenty-eight. A little matter now and something else that would not concern the casual audience to whom this article seems directed. Holland interrupts the story of Barringer's work at the crater to describe Barringer and his early career. Holland writes, "Incidentally I may mention that Mr. Barringer, a highly cultured Southern gentleman, and the late John Brockman, a desert prospector, had been partners in the Commonwealth Gold Mine, Arizona, and had made nice stakes. Brockman would have nothing to do with the Crater, but put his money into Los Angeles real estate and died a millionaire, the Brockman Building at Seventh and Grand being part of the estate. Far be it from me, however, to encourage anybody with spare cash to buy Los Angeles real estate instead of drilling wildcats." Surely an audience of oil drillers found Holland's last statement humorous. In truth, Barringer had another partner besides John Brockman named R. A. F. Penrose. These three men actually only controlled the Commonwealth Mine until 1910 when it was purchased by the Montana Tonopah Mines Company. The high-grade ore had mostly played out by 1905 and that same year the shaft collapsed.

The Commonwealth Mine in Pearce, Arizona had been one of the largest strikes of silver ever made up to that time. The mine had produced considerable gold as well, but where Holland has referred to it as a gold mine in the above quote most often, it would be considered a silver mine. There is little doubt that it was the early source of Barringer's wealth. We often hear stories of prospectors striking it rich, then going on to lose what they have earned and die broke. John Brockman according to Holland chose wisely and left the world of prospecting and grubstaking, moving to the big city to become very wealthy. But, it is elsewhere recorded that John Brockman was a Silver City banker and purchased the claim from John "Jimmy" Pearce for $250,000. He had to demonstrate that he could get the claim operating in 90 days to complete the deal. He fulfilled that obligation in only 60 days. The Commonwealth produced 8 million dollars of silver at only 50 cents an ounce and 2.5 million dollars of gold at $20 an ounce. Other estimates put the weight of the silver produced at 12 million ounces and the gold at 138,000 ounces.

Holland's article becomes personal at this point as he writes about his own work at Meteor Crater. United States Smelting Refining and Mining Company had decided there was to be no more drilling into the crater floor. "But to drill on the rim in the hope of getting on the course taken by the possible shower of meteorites which may have become buried somewhere outside the Crater's circumference. We received the advice of noted scientists and some of my happiest recollections in connection with the Crater are the personal visits and letters I received from famous astronomers and physicists."

There is no doubt that Holland was an accomplished mining engineer but not any more so than hundreds of others. He was thrust into the world of truly famous personalities. He had letter, and telegram correspondence with Professor Magie a world-famous physicist and Dean of Princeton University, Elihu Thomson the President of MIT and one of America's most famous inventors and industrialist, and the list could go on. It must have been a thrill to have these men come to the Crater and to be their host and guide. To have the opportunity to talk with them and exchange information. Holland was the man solely in charge of the crater, and it must have been a great boost to his ego having these men look to him for aid during their visits. Barringer and his sons were also visitors, and it is clear from his earlier statement that Holland held him in high regard. And Mr. Barringer was certainly very friendly to Mr. Holland during his visits.

The article returns to Holland's work at the crater by discussing the elaborate magnetic survey and the nature of the arch of the south rim with its creviced and broken layers of uplifted rock. Holland is now ready to begin discussing the actual drilling. He writes a caption for a photograph taken of the work site from the west showing the drill derrick. This is not the first mention of a photograph being part of the article. Despite all the humor and personal anecdotes up to this point, this caption could indicate the possibility the article was for print.

"For the first 300 feet the hole penetrated a very hard limestone, so badly creviced that the cable tools were continually in trouble. As many as four stems were broken in a week. Fishing Jobs were the usual thing rather than the exception. More than once it appeared that about all the equipment except the kitchen stove was in the hole. Needless to say, no drilling records were broken, though other records are perhaps worthy of note. At a depth of 200 feet, the bailer brought up a rat's nest. At a depth of 320 feet, where the limestone was in contact with soft sand, there was so much junk in the hole that it could not be sidetracked or drilled up. To have skidded the rig, as was first proposed, would have meant again going through the grief of drilling the creviced limestone, so the drilling crew were given a rest while a small tunnel 400 feet long was driven from inside the Crater to the bottom of the hole to clean it out – something of a curiosity in the way of fishing jobs." Holland has entertainingly compressed the horrors of the drilling operation into the last quote. But his repeated reference in other documents that the work was "hell and even worse for him personally" is probably a more accurate depiction of his feelings. The contracted money to be spent was nearly used up by the time the drilling began. He had spent months preparing the site, having the camp built, and getting the pipeline laid. It may be that after several years he was able to look back at that time with a little less pain and write as he has with some humor.

Holland now begins relating the facts as he knows them about what happened after he resigned from Crater Mining Company. "After the hole had been cleaned out and drilling resumed in the sandstone, a rotary was substituted for the cable tools, but became irretrievably stuck at a depth of 1376 feet, where water ran freely out of the bottom. More or less iron oxide, carrying nickel, was brought up with sandstone fragments in the lower part of the hole, but no metallic meteoric iron. In a published paper, Mr. Barringer maintains that the drill hole located a meteoric mass, and pointed the way for future exploration." The use of oil drilling jargon such as "rotary" and "cable tools" again leads this writer to think Holland is addressing oil industry workers. The drill did become wedged in place at 1376 feet and could not be freed ending the operation and further drilling by USSR&M Co. However, the drilling log of Superintendent Plumb does record that the bit hit two spots that were only a few feet thick but so hard that it took days for the drill to grind through the material. It was as hard nearly as the bit. The nickel tests were very good at those spots as well, and it is quite possible that two metallic fragments of the asteroid were actually discovered by the drilling. However, the material that Barringer calls a meteoric mass "discovered" by the drill near 1350 feet was always a mixture of various rocks along with the meteoric iron shale. There was never an actual continuous mass of weathered asteroid drilled through, just about thirty feet that had a larger amount of the iron shale mixed in with other rock. The mention of the water running out the bottom of the hole is important. It means that there was another large crevice and that they had the problems associated with trying to drill a dry hole. This would have been very difficult or impossible to do with the churn drill cable equipment. They would have had to seal the bottom of the hole by dumping material into it from the surface so it could hold water. Holland and the drillers had to do this on occasion when crevices were letting the water run away. It was undoubtedly difficult with the rotary equipment to drill without water also. Perhaps the most important aspect of this statement about the "water running freely out the bottom" is that it might confirm Barringer's belief that the water which had flooded the shafts on the crater floor was confined to the crater interior. The dry hole surely confirmed it in Barringer's mind. It likely contributed to Barringer's later attempt to reach the same area with a shaft and tunnel from the south flank of the crater. At 1376 feet a drill hole was as much as 600 feet below the start of the quicksand discovered in Barringer's shafts on the crater floor. It would seem safe to dig outside the crater to a similar depth. Barringer's dream remained even after all the problems and costs of the drilling program on the south rim. In the final attempt to tunnel over to the buried mass, he hit water outside the crater at just 600 feet. He would with great difficulty and expense struggle down to a little over 700 feet but finally lose again to the water. This detail about the hole running dry at the 1376 foot depth is stated by Holland. It is used by Barringer and has been quoted in works elsewhere, however it is not mentioned in the foot by foot notes of Superintendent Plumb on his drill log. It is possible that it is true and was reported in Plumb's weekly reports that he undoubted made as Holland had to the home office. Barringer would be copied on these as he had with Holland's and perhaps had that added knowledge.

Holland without any judging writes the following about Barringer's final exploration. "The most feasible method of proceeding appeared to be the sinking of a shaft, at a probable cost of a quarter of a million dollars. Considering the hardness of the material, it would not take a very big iron meteorite to put any kind of a drill out of business However, taking into account the difficulties and cost of further exploration by shaft sinking or drilling, the majority of the directors of the Crater Mining Company, (the name we operated under) decided not to continue operations and turned back the lease to Mr. Barringer."

Just some simple math indicates that the water inside the crater and outside was close to the same level and that the crater was not a confined basin of trapped water. Barringer struck water and quicksand at 180 feet in his first shafts. The average increase in the crater rim height from that of the surrounding plain is 157 feet. The crater is approximately 570 feet deep. If we subtract from that depth the 157 feet of the rim, it leaves 413 feet that the miner must dig through to be at the same level as the crater floor. Add to this the 180 feet that the Barringer miners dug until they hit the water and the depth is 593 feet. The water outside the crater was struck at nearly exactly 600 feet. This is a simplistic view because the final shaft was far down the slope and already well down from the top of the 157-foot average rise of the rim. Also, the numbers I used are averages. Still, it is clear that the water table was at similar depth inside and outside the crater. Exploratory work was done prior to Barringer's final flooded shaft on the south slope. But apparently, no one did even simple math or drill down fully to 600-700 feet and see if there was water. Barringer did continue to stalk the asteroid. In 1928 he created yet another company to raise money for this work on the south slope. That final Barringer venture ended up being the most expensive of all the explorations.

This concluded the first part of Mr. Holland's article. Part Two takes a different look at the crater and the work.

The scientific community had been debating the origin of the crater since Grove Karl Gilbert's first visit and his declaration that a volcanic steam explosion formed it.

Scientists such as George Merrill wrote and presented evidence that it was created by an asteroid falling out of space and striking the Earth with tremendous force. The debate escalated in the 1920's to a real battle of opinions. Finally, mathematics would begin to rein in some of these wild ideas and establish some boundaries for the possibilities. It was becoming clearer that only an impact from space could have created the crater. But, how big was the asteroid? Was it millions of tons traveling slowly as Barringer thought or was it much smaller with a high velocity? Did the asteroid survive or was it consumed and vaporized by the impact?

Holland begins Part Two by stepping into the arena and taking up the impact origin theory. Just what one would expect. He had actually spent more time personally at the crater than nearly anyone else alive. It is very likely that he was there, even more days than Barringer's total. He saw the rocks and the meteorites and as Barringer had always said the Crater was able to tell its story better than any person could. Here is what he writes.

"Many opinions have been expressed as to the origin of Meteor Crater, described in our last issue, but it may be said at once that eminent scientists whose opinions carry the greatest weight believe that the hole was caused by a shower of meteorites and the accompanying gases and steam which would be generated when the mass passed through the earth's atmosphere at terrific speed. These gentlemen include such authorities as Professor Elihu Thomson; Dr. W. W. Campbell, President of the University of California and former Director of Lick Observatory; Dr. W. F. Magie, Dean of Princeton University: and Dr. George P. Merrill, Head Curator of the U. S. National Museum and perhaps the greatest authority on meteorites in the world."

Mr. Holland had corresponded with Thomson. Dr. Magie visited the crater and Holland was his host, they were both strong supporters well known to Holland. It is Dr. Campbell which sticks out in this list. He visited the crater but was not publicly much of a vocal supporter of Barringer's ideas. They had a bit of a disagreement over when Campbell could visit the crater and Holland received instructions to treat him rather coolly, letting him look around on his own, and not to give him too much information. Yet, Holland includes Campbell in his list of respected scientists who favor the impact origin of the crater. It must be wondered if Holland did take some personal time with Dr. Campbell and his wife when they visited the crater. Did Holland get to know the man and learn that he was, in fact, a strong believer in the asteroid impact theory? Campbell supported Moulton's impact theory of a small size body with high-velocity. These beliefs made Campbell a less valuable supporter to Barringer. Mr. Holland may have continued to be interested after he left the crater, keeping himself informed as to the progress of the debate.

George P. Merrill had written a comprehensive paper on the crater in the early years just after Barringer and Tilghman reported on it and prior to Barringer's second paper on the Crater. Merrill's paper was in direct conflict with that of Gilbert who was Head of the U. S. Geological Survey. But the National Museum was a different agency of the government and Merrill felt free to say what he knew and had learned. By the time that Holland came to the crater Merrill's paper would have been already old. But Merrill produced a supplementary report within the time frame of Holland's work at the crater and the writing of his "Drilling for Meteorites" article. For him to know says he did some research on the crater himself or perhaps the paper was made available to him during his work at the crater. A copy of Merrill's first paper could easily have been in the old stone museum where Barringer stored rock samples for visiting scientists to study. Barringer was often in communication by letter with Merrill.

Holland continues, "How much of the damage was caused by the meteorites themselves and how much by the accompanying gases and steam it is impossible to estimate. The earth is constantly being bombarded by thousands of meteorites, a very small proportion of which ever succeed in surviving their passage through the atmosphere. A meteorite has been known to fall on a frozen lake and to rebound without breaking the ice. Few have actually been known to penetrate more than a few feet into the earth. Although meteors enter the atmosphere with velocities varying from 5 to 45 miles per second, their speed becomes much more moderate after they encounter the resistance of the air. They become very hot on their first entrance into the atmosphere, so that we see "shooting stars" (and take the occasion to wish that Calpet stock will go up,) but they are cooled again in the lower air, if they survive, and are often merely a little warm to the touch on reaching the earth as meteorites. Such a cataclysm as would cause the Meteor Crater, and melt sandstone, is outside of authentic historical experience."

Holland does a quite acceptable job of describing the fall of a meteoroid from space and the landing of the few which survive to become meteorites. Since the time of his writing many more meteorites have been recovered, and much of what he has stated has been further studied. It has been told throughout history that meteorites have landed and were too hot to handle sometimes even starting fires or charring the grass upon which they land. This has been a commonly held belief, but Holland does not present any of that dramatic legendary material here. He relates some good science and what is still seen today. Most small cosmic bodies fall with a terminal velocity of about 200 miles per hour after they stop being luminous. Several famous meteorite falls have taken place on frozen lakes since Holland wrote his article and what he wrote has been true. The meteorites are found lying on the ice surface their black fusion crusted exteriors easily seen from a distance. Meteorites according to the math should land cool or at most slightly warm as Holland writes, yet the stories of them being too hot to handle continue to be told. Two recent mathematical studies on this topic have revealed again that they should not be hot. It was not the heat of the asteroid's passage through the atmosphere which caused the explosion and excavation of Meteor Crater. It was the compression of the asteroid and the rocks at the impact site which caused its formation. The creation of a crater by asteroid impact is thankfully still outside our historical experience. But we have created similar craters by exploding atomic bombs underground. The energy release from the two events is so similar that much can be learned about asteroid impacts by examining nuclear bomb craters. Over the last few decades, while monitoring the planet for atomic explosions, sensors have detected many explosions of meteoroids in the atmosphere. They are often events with atomic bomb force. The Chelyabinsk Meteorite event in 2013 yielded one-half megaton of energy. Fortunately, the explosion was high in the atmosphere otherwise, there would have been many more injuries and likely deaths connected to the event. It was serious enough; with over 1500 individuals injured mostly from flying pieces of broken glass.

Holland inserts the strange parenthetical remark about the Calpet stock going up in value. Again a suggestion this is being read to an audience of oil workers in California. If it was such a speech, then it was a presentation with slides or some other method to show the pictures he describes.

The article takes another direction as Mr. Holland begins a discussion of the Native American oral history about the crater and the early scientific studies which created the debate which was still raging on. "The Indian tradition is that over a thousand years ago one of their gods came from the sky and buried himself in the ground. The legend says that every living thing in the surrounding country was killed except the people who were in the cliff dwellings. The Hopi Indians still use white flour rock from the Crater as a sacred substance in their ceremonial dances. One Indian told me that the god in the Crater would object to our disturbing him and that we would have no success. It may be noted that some of the stunted juniper trees on the rim at the south end of the Crater are over eight hundred years old, so the damage must have been done at some time prior to this period, as the trees certainly could not have survived the catastrophe. The Indian story is not one that they would be likely to invent as it is quite outside of their experience and knowledge. The idea of volcanic activity and an internal explosion would have come much more easily and naturally to them in explaining the origin of the Crater."

Meteor Crater was reported by H. H. Nininger to be a taboo location for the Native Americans he had contact with when his meteorite museum was on Route 66 near the crater. He occupied the towered building that is now in ruins but visible from I-40. The tribal elders told younger tribe members that they have nothing to do with the crater or the rocks found there. The elders kept the younger members from going into his museum. Holland reports there is folklore among the Native Americans that is similar to the real events that we know transpired there. However, the present recognized date for the impact event is just under 50,000 years ago. There appears to have been no human presence in North America at that time. Stories told generation after generation can last within a culture for a very long time. Studies done on the survival of cognitives in language also show that special classes of words can withstand change for dozens of generations. It is necessary to look at Holland's report of the tribal stories critically. While it is unlikely that Holland made them up, it is possible that they are a more recent creation of the tribe. Or the tribal people with their close connection to, and reverence for the land had some understanding of the geology that led to the creation of a story that fits the uniqueness of the Crater's form. They may have had other information to make a connection between the strange rocks and things falling from space. Within the same time frame as Holland's article, the late 1920's two ceremonially buried meteorites were found in archeological ruins within a few miles of Meteor Crater. Both were buried similar to the way a small child would have been buried wrapped in a feather-cloth and placed in a stone-lined pit. In the case of the Winona meteorite, the earliest publications of the discovery suggested that the meteorite was buried where it fell and that the ancient people may have witnessed the fall. In the case of the Camp Verde Meteorite, it was later determined to be a transported Canyon Diablo iron from Meteor Crater. The night sky for Arizona's ancient Sinagua people was the realm where their deities lived. Great importance was attached to shooting stars, comets, eclipses and bright fireballs of course. Were they able to make a connection to the iron meteorite fragments at Meteor Crater and the other meteorites that occasionally fell which they ritually buried? The Camp Verde Meteorite weighing 135 lbs. (61.5 kgs) was removed from Meteor Crater and transported to the top of a mesa and likewise wrapped in a feather-cloth and buried in a stone-lined pit about 800 years before it was discovered by a relic hunter.

Holland's use in his article of information about the scrub juniper trees is something that was reported early in the investigations by Standard Iron, Barringer's first company. Samuel Holsinger who had first told Barringer of the crater's iron had seen the woodpile at Volz Indian Trading Post. Mr. Volz had cut the wood from the crater's slopes. Holsinger took a few sticks and counted the annular rings. He obtained an age for the trees of over 500 years which he reported to Barringer. Upon looking at the trees on the crater's slopes, Holsinger determined that many were even older, six to seven hundred years of age. The eight hundred years used by Holland in his article is a greater age but not that far out of line. However, by the 1920s many of the scientists were even then thinking is terms of thousands of years for the crater's age. This increasing age was based on evidence such as the corrosion of the limestone since the impact occurred, and the altered condition of the buried iron masses that had totally deteriorated into shaleballs.

Holland continues: "One turns to the reports of the U. S. Geological Survey for something more authoritative than the Indian tradition - and is disappointed. Here we find the hole described as a "volcanic crater" though there are no volcanic or igneous rocks nearer than nine miles. No doubt a very cursory and hurried examination of the locality had been made, and Mr. G. K. Gilbert, of the Survey, in an interesting paper published in 1895, corrected this error, but not in the Survey publications. He substituted the theory of a steam explosion for the volcanic origin of the Crater, but this has been shot to pieces by Dr. Magie and other scientists."

Holland is a little off base here. The investigation by Grove Karl Gilbert had not been "cursory" but exhaustive. Gilbert spent seventeen days at the crater. They took photographs and did a magnetic survey. He calculated the volume of the crater and the volume of the material thrown out of the hole, and he collected many rock and meteorite samples. It will always be true that he declared the crater the result of volcanic activity which of course the steam explosion also was. The reason for this declaration was he did not find the information he was looking for to prove a cosmic origin. Gilbert had left Washington D. C. with the belief that the crater might have been made by an asteroid. It was his silence and position of authority that helped to keep the debate going. Had he ever truly corrected himself once it was clear the crater was not volcanic he could have cooled the argument considerably.

Holland takes up a bit of that raging debate in the next portion of his article. "An entertaining controversy recently took place in the Engineering & Mining Journal between Mr. Dorsey Hagar, Oil Geologist, and Mr. D. M. Barringer, Mining Geologist I am personally acquainted with both gentlemen and will let them argue in their own words. Mr. Hagar wrote, in part:-

"The so-called Meteor Crater, so far as I can see, is a big hole in the heart of a perfect elliptical shaped dome in sedimentary rocks, limestone, sandstone and shale. The hole in the heart of this dome can be explained as due to solution of underlying gypsum, salt beds, and limestone beds. So far as I can see there is no evidence pointing to either volcanic action or to meteoric action. Meteoric fragments are quite common in that part of Arizona and not confined to Meteor Crater."

"Mr. Barringer's reply is too long to quote in full but he said, in part:-"

"At last I have seen the light, and the origin of the Crater is now clear. It is a sink hole. But not an ordinary sink hole such as is often found in limestone. Far from it. This sink hole was found in sandstone (97.5% silica) and what is more, instead of sinking, the rocks in this remarkable hole rose up, and fell outside of the hole, some of them even flying more than a mile away from it. And at the same time some of the rocks in the hole were highly heated, some of the sandstone fragments so highly that they melted - as no other sink hole has so far been able to cause. And millions of tons of sandstone, before coming out, were shattered into finest flour rock - also, no doubt, by sinking. It gives one pause to think how easy it was for Mr. Hager to figure all this out in one visit, and to put it all clearly and conveniently in a half column letter. Oil geology is manifestly too small a field for him. To think that I have continually been encountering meteorites in Arizona without recognizing them! It is obvious that we mining geologists have a great deal to learn from some of our petroleosophical colleagues. Let me ask Mr. Hager, if he has any further information about the crater, not to withhold it any longer, for it is against the highest duty of the scientist to retard, in any way the spread of scientific knowledge."

Of all the theories thought up for Meteor Crater Mr. Hager's is one of the poorest. He must not have seen anything when he was at the crater for his one visit. As Barringer states, the crater foundation rocks are sandstones. The Coconino Sandstone is a layer approximately 800 feet thick. It is only exposed around the crater walls for the top two hundred feet. There is more sandstone of a different type referred to as the "red beds" below the Coconino. There is, therefore, nothing to dissolve away to form a sinkhole. Dissolved Kaibab Limestone is not the material on the crater floor. And it has not dissolved away around the rim. The limestone has as Barringer says in his rebuttal to the magazine article been thrown out around the hole. The boulders on the east and west sides of the crater are great blocks of limestone weighing often hundreds of tons. Barringer's sarcasm is uncharacteristic of the gentlemen he usually is. But this was such a badly thought out theory created by a person with no knowledge of meteorites or impacts that perhaps Barringer can be forgiven for his sarcasm. It is interesting that with all the evidence in Holland's article that his audience is the oil industry, that he would put a member of that community up for ridicule. Especially after his admission that he is personally acquainted with Mr. Hager.

The understanding and acceptance that meteorites are from outer space took hundreds of years to be realized. Holland moves on to a discussion about the resistance some in science always seem to have to new ideas. At the time of his writing, Meteor Crater was going through a struggle for recognition as the first impact crater ever found on Earth. Here is how Holland presents the controversy to his audience. "During the whole of the eighteenth century, the Academie des Sciences of France, the foremost scientific body in the world at the time, persistently maintained that it was physically impossible for bodies to fall from the skies to the earth, and they ridiculed the credulity and superstitions of those persons who claimed to have seen such falls. This is a solemn warning for all time to any person, scientific or otherwise, who ventures to give a final verdict upon matters outside of his immediate experience. Nevertheless, after looking at the crater every day for a year, I have the temerity to offer the opinion that there is a direct connection between the meteorites and the hole as cause and effect, whatever may be the possible commercial advantage of digging in the ground for some of the cause. As a sporting proposition, the betting against a mere accidental coincidence of ninety percent of the world's known stock of iron meteorites falling at the exact locality of the Crater is many millions to one."

The evidence for the origin of Meteor Crater was multifaceted. There was the physical proof in the rocks and the way they were displaced, heated and pulverized. There was as Holland writes the vast number of meteorite fragments lying around the hole. Was their presence in that location by coincidence or from the event which created the crater? The region of Arizona where the crater is located is a volcanic area. Many volcanic peaks are in view standing on the crater's rim. Could the work done at the crater in excavating the pit and throwing the rock for as much as a mile be measured? With all the heating and melting seen in the rocks could the asteroid itself have survived? How fast and how big would an asteroid have to be to do the work seen? Size and speed are part of the same mathematical problem. Where is the evidence of either item to make the calculations? It seems that it was after Holland wrote this article that Forest Ray Moulton one of America's premier mathematicians took up the challenge of determining if the asteroid could have survived and what the size of the impactor might have been. Even for Moulton, it was a process that required several attempts to get an answer that was hard not to accept. But in 1929 with the release of his last report, the results were clear there was no asteroid of millions of tons buried anywhere. It had been much smaller and much of its mass without doubt vaporized during the impact. By this time Barringer and new partners were deeply ensnared in another futile attempt to reach the buried iron. They were trying to dig the shaft that Holland mentioned near the beginning of this article. They planned to dig to the depth of the shale ball iron found by the drilling program Holland began. Then to tunnel horizontally to the location. The mouth of this final shaft is still out on the south slope a thousand feet from the deserted 100 foot by 40 foot leveled drill site that Holland created. If nothing else the final shaft is the last fenced off testimony to the unshakeable belief of one man; D. M. Barringer.

This photograph is of the still open hole of the final shaft near the foot of the south slope. It is about 1000 feet from Mr. Holland's south rim drill site. Abandoned like many of the others when it struck water at 600 feet. With ever larger pumps and by cementing up the walls the men were able to push their way down to just over 700 feet. When the pump failed the water flooded the bottom hundred feet of the shaft, and the project was abandoned. It was the final attempt to reach the iron mass that Barringer believed the drill on the south rim had located. Nearby are the foundations that great machines were mounted to. The wenches and pumps are gone but the concrete foundations remain.

Holland's "Drilling for Meteorites" was preserved in his papers in four versions. Most end at the quote above about the odds of the crater forming where the world supply of iron meteorites also fell. But one of the versions adds the story of a study at the crater to which Holland became aware. It involved a newly invented machine called the "Radio Cameraphone." Holland gives a long description of the machine's mechanism and its use at the crater. Here is that last section from his final version. They are also probably the last words that Holland wrote concerning Meteor Crater and his year of work there.

"An interesting account by Wm. A. Sharpe, of a recent attempt to locate the meteoric body by means of an instrument called the radio cameraphone, has just been published in the Mining Congress Journal. This instrument was developed by Mr. Sharpe from a similar appliance he has used in experiments in Colorado and New Mexico, to locate oil and gas pools. It should not be confused with the so called "doodle-bugs." Its experimental use is endorsed by engineers of high standing, including D. W. Brunton, the inventor of the "pocket transit" which most engineers carry nowadays. Mr. Sharpe describes the instrument as consisting of three units, the radio broadcaster, the radio receiver and the photographic motion picture camera. The broadcaster sends radio waves into the earth, extending to regulated progressive depths, the length of the wave advancing one foot at a time. This radio wave is returned by a closed circuit to the receiver, where the tone pitch of the sound varies according to the mineral or substance at the far terminal of the wave. Recognizable variations in tone are produced by different minerals and metals. The photographic unit is fitted with a vibratory illuminated thin metallic foil about one inch square, suspended by a fine hair spring which is in contact with the diaphragm of the telephone mechanism in the radio receiver. The illumination of this foil is reflected in a magnifying mirror which intensifies it 30 or 40 times. The reflected light of the foil in the magnifying mirror is thrown on a ground glass screen, the vibrations appearing as lights and shadows, and these are photographed by a regular motion picture camera, which is automatically exposed. The photographs may then be thrown on a screen by a regular motion picture projecting machine."

"Mr. Sharpe states that the photographs obtained at the Crater show the main unoxidized body of the meteor to lie 100 feet south of the rim of the Crater, the center of the body being at a depth of 1410'. If the instrument tells the truth about the position of the meteoric body, it indicates that with some miles of the circumference of the rim to choose from, we located the Crater Mining Company's drill hole just right, though the radio cameraphone had not then been invented."

"As previously mentioned, the location of our drill hole on the rim was decided on as the result of study of the physical condition of the strata, confirmed by a magnetic survey which indicated a line of deviation from the normal magnetic variation of the district as passing through the site chosen."

Thus ends Mr. Holland's article or speech. His life moves on it seems with no further mention of Meteor Crater. He receives a single inquiry about the crater years later which he places with the rest of the saved documents from his year of work. We can only guess whether he thought about the crater in the decades to come.

After the death of D. M. Barringer, many new machines and methods were employed in more attempts to find where anomalies occurred around the crater. All the instruments recorded things that the operator said were a buried objects. The areas of these "hits" were never in the same spot. Some had the body within the crater others outside. The tools of the 1930s to 1950s were just not much better at finding any evidence of buried iron than the original dip needle that Gilbert employed in 1891. Aerial Magnetometer studies even later showed nothing at the Crater. And that is perhaps what should be indicated since we now believe that most of the asteroid was turn into nickel-iron vapor which later condensed into the tiny droplets found in the soil.

Barringer would not get the credit that he deserved for the discoveries he made at Meteor Crater. Much of the scientific community would keep silent or cling to the volcanic theory even after Merrill papers and Moulton's reports and Barringer's evidence. Decades would pass until finally, some organizations would give the credit to others in the 1950s and 1960s. These individuals would repeat the work of Barringer. And though they had read and knew what Barringer had done they would sometimes publish materials as if it was all being learned for the first time. Today the crater is called within the meteoritical community "Barringer Meteorite Crater" finally acknowledging the pioneering work of Daniel Moreau Barringer Jr. Mr. Holland and the drilling crew, on the other hand, disappeared completely into the realm of forgotten history.

The Aftermath

What Remains

Well, that is it; we're done. 1,376 feet and boulders are wedged behind the tools. It has been like that for weeks now. Nothing we have tried has worked to free the bit and get drilling again. The word came down today that we are to disassemble the derrick. We are to pack up and ship the salvageable materials. Send it all off to the railroad station. The wood from the derrick and rig I guess we will push that over the edge to get it off the drill site.

I do not know for sure if we found anything. Barringer seems to have mixed feelings. One moment encouraged and then more questions as if he has doubts. We did hit some meteoritic material down around 1,300 feet. But not much of it was metallic. Just those two spots we could barely get through. Grinding away and bit resharpening for days at whatever it was. We were only able to drill down a few inches during a whole day's work. And we got strong nickel tests in places. But soon we were back in sandstone and no nickel. Whatever we hit was not one big object but more of the alternating layers of sandstone, rubble and small masses of iron shale with of course those two larger somethings mixed in. I think they could have been iron meteorites.

Now we are just going to throw it all over the edge and disappear to other work where ever we can find it. United States Smelting has been cutting back on its exploration work. That is how I got here. They let Holland go so I could stay employed. I have been with them for many years. But I never had so many problems anywhere as I have here. I know they can not be very happy. This work cost them far more than they had planned or promised. They may let me go too.

I have thought a few times about Holland. I had the same problems he had, and I will likely be criticized the same way I guess. The Barringers have sure bad mouthed him since he left. They smile and appear understanding when they are here for visits. The men in Boston read my reports and never say anything bad, they did that with Holland too. But it is always the men who actually work and suffer who get blamed when other people's dreams don't pan out. It is always the bosses that get patted on the back when the mother lode is found. It is the pecking order of the world I suppose. Holland and I did what we were asked the best we could. In the end, we will both be called names and talked down about. In a hundred years will anyone even remember the names Holland and Plumb. It was this place and the mystery of what happened here thousands of years ago that caused the struggles and failures. I wonder if they will ever find anything left from the asteroid. Maybe it just blew apart like the artillery shells in the war.

So tomorrow we begin taking the rig apart and start pushing all the wood over into the crater. There it will stay for years to come. A visible reminder of our work and perhaps of our failure. But at least as long as it stays strewn about the wall and floor of the crater no one will be able to forget that Holland and I were here with a group of hearty men trying to find a fallen star.

The drill site is cleaned up some today. A few timbers and lengths of pipe remain but if you look a bit harder you will see much more is still there. Then you may notice a brick here and a piece of metal there. Parts of machinery buried just under the surface are exposed when you push the sand aside with your shoe. The flat platform dug and blasted from the crater rocks will remain much as it is for hundreds if not thousands of years. From 4000 feet away at the museum on the other side, you can not see with the naked eye the remains of the drilling program of 1920-1922. But if you use the provided telescopes or even a telephoto lens you can see the hundreds of pieces of lumber that were once a drill rig and derrick. It was all once on the top of the south rim. You can see if you look at the top of the cliff the flat groove cut nearly to the bottom of the red sandstone by Holland's laborers. It is now an unnatural sharp dip in the gently rolling surface of the crater rim. The upper half of the red Moencopi very obviously missing. A post still stands on the top of the knoll near the drill site. It was likely where Barringer had the flag placed marking the southern end of his MN line from the first magnetic study.

One of the great beams of wood thrown over the edge is resting nearly on end. It has a notch cut into it showing where cross beams were bolted to it. Some pieces of the derrick are shattered from falling hundreds of feet. Many repose in bunches where they slid to the same location, flowing to those spots along ridges and valleys in the talus.

Running through the pile of timbers is the red streak of the crater. It was never there before the drilling. The sludge from the drill hole was dumped over the crater edge. Originally the mud was white when it flowed down the face of the south cliff like a bright waterfall. Over time it has rusted from the iron it contained. The years have washed away nearly all of the dried mud. Eventually, erosion will remove the red debris completely from the talus. Undoubtedly the red stained rock layers of the south cliff will be colored for thousands of years.

Roads are some of the most tenacious scars on a landscape. And with the desert climate of Meteor Crater, the roads are still easy to see on satellite images. They will likely be visible for hundreds of years with imaging systems created in the future. As space-based tools are used to see the buried cities of the ancient Egyptians the roads at Meteor Crater will continue to be seen.

The remaining camp buildings will fall apart. They are barely holding together today. They will collapse, as others have done. A strong wind will flatten them sometime in the future. The nails will remain for decades. Someday they will become just more iron particles in the soil to mix with the meteoritic bits already found everywhere around the crater. Brass, bronze, and copper in the form of screws, wire, plumbing, machinery parts, bullet casings, and lost coins will last for hundreds of years, some until the end of the world. Just as the sunken bronze parts of the Titanic shall remain after the iron hull is gone.

The bricks at the drill site were made by the Los Angeles Pressed Brick Co. a business which is gone. The bricks will remain forever as long as they are not crushed. Perhaps they will be found in the distant future by archeologists who will not know what the letters molded into their surface mean. The concrete foundation for the gasoline engine is not visible now but maybe it was broken up, and the pieces remain buried somewhere awaiting excavation. Or perhaps it was pushed into the deep cellar that had been blasted out under the drill rig. The concrete foundations at other locations around the crater remain. They will last for thousands of years, but likely become broken up into chunks as the rebar steel rusts and swells splitting the concrete.

The crater is 50,000 years old and has survived the worse assault upon it; that of man. If our race leaves it alone, it may survive for thousands more. It will heal naturally until all but a few traces of the work of man are gone. The meteorites in the ground around the crater are being left where they are. Some complain this is a poor idea, that they are just rusting away. But nature has coated the iron meteorites with magnetite and a thick layer of iron shale. Not much will change with them over the next few hundred years, and it is good to leave some for future study in place just where they fell. Those buried under the crater floor seem to have suffered worse and are mostly masses of iron shale. That seems to be what is buried under the south rim also. The thousands of years that Meteor Crater was a lake has taken its toll on most of the asteroid fragments within the crater. The drill casing in Barringer's 28 holes will disintegrate, and the shafts will fill with sand and debris. The central white spot made of shaft tailings may remain for hundreds of years, but it too will someday be covered or blown away.

All the things that man has done to the crater will disappear except for the concrete and brick buildings of the modern museum and the stonework of some of the old buildings. The scars made to the crater itself will fade. It took just thirty years for most of the wounds to be inflicted, but it may take thousands of years for the evidence of man to disappear. Ancient man was not there when the crater formed. Will man be there when it finally heals? Will we be as interested in Meteor Crater once we move out into space and have millions of similar and far larger craters to explore? Will Meteor Crater fall into neglect, unvisited because of disinterest? Or will there always be a few who cannot resist walking its rim and feeling the soft spongy soil of its floor? Like the desert explorers of old ghost towns today maybe a few souls will still seek out Meteor Crater in the distant future and fall under the enchantment of its silence.

The following photograph shows the few remains of the original stone museum which Barringer had built and where he stored the collection of rocks for visiting scientists to study. It was from this building that Holland retrieved the shotgun, pistol and cartridges on the morning of June 30, 1920 as instructed in Barringer's note. The building held the Holsinger Canyon Diablo meteorite also. It accidentally burned down in the late 1950s when some coyboys took refuge there during a storm. So the story is told.

L.F.S. Holland After Meteor Crater

I sent Lillian a wire when I was in Winslow yesterday; she will meet me at the train station in Los Angeles. Taking Plumb into town to introduce him to all the people we work with and giving the bank his letter of introduction, were the last of my duties here at Meteor Crater. As of this morning, I am no longer an employee of Crater Mining Company or actually United States Smelting Refining and Mining Company. They were supposed to send my last paycheck so that I would have it before today. Now I have to wire Boston to find out where it is when I get home.

This job has been hell for everyone and for me personally even worse. I wish that I could have found the iron meteorite. I think it could have been a historic event. Maybe Superintendent Plumb will have better luck. But I do not believe in my heart of hearts that he will. We made very few real mistakes. It is the crater itself that is the problem. He will have to finish the tunnel first. We got to 203 feet just barely halfway. If the rock continues the same, it's going to have spots that are very bad. I thought we were going to lose some men when the timbers and spiling cracked and again during that one long run of sand. I dreamed about the shafthouse collapsing again last night. It was not a nightmare but a very vivid dream. I guess it was providential that we were out of it a few hours before it fell.

I would like to get Lillian a little something from Volz's Indian Trading Post before getting on the train. But do not know what she would like. It is a few miles further to Canyon Diablo than Sunshine Station. Maybe I will see something at Canyon Diablo if I can get them to take me that far. Otherwise, there is a stop at Flagstaff with enough time to get off and look around for something. Maybe a piece of Indian turquoise and silver jewelry.

One more walk around the camp and drill site to say a few goodbyes. Especially to the cook. He was the best cook I ever had at a mining camp and worthy of the praise I had heard before I hired him. I took a few more pictures. Been working on an article about the job here. "Drilling for Meteorites" is what I am calling it. . .

This Overland Automobile has done as well as a cheap car could, and the Ford truck with the special gearing hauled almost everything. They usually got us where we had to go, but I will certainly not miss these terrible roads. We are almost to Canyon Diablo Station for the last time. Two days and I will be home in Hollywood. I'll have to run over to the office of The Mining Journal or call them to let them know I am back in town. They can put an announcement in the next issue. It has been hard living at the crater. No telephone, no entertainment, very cold and windy, plus plenty of problems. Jennings in Boston said he would be happy to give me a reference and help to find me another position. I think he and Anderson understood the difficulties. I will always feel that if the gasoline engine had not failed, then we would not have stopped for those ten days. The casing would not have gotten cemented in place by the lime. We would not have shot off the shoe and the bottom of the casing. We would maybe not have had the underreamer and 8¼ inch bit in the bottom of the hole. Almost certainly the underreamer would not have been broken in two. But it is all water under the bridge as the Americans say. I have a couple of prospects for work in Mexico. It is just exploration work, but the mining industry is growing there. It could lead to something else. There is plenty of work in Los Angeles too in the oil fields. Have to try and keep a positive attitude this is not the first time I have been between jobs and likely will not be the last, and we have set aside a good portion of the money I made here.

The last official piece of paperwork that L. F. S. Holland saved in his files was the response from the assistant treasurer of United States Smelting Refining and Mining Exploration Company about his final paycheck that he did not receive by the beginning of May 1921. His employment ended as of May first, and in the letter from the company to him, he is assured that the paycheck was sent and was to arrive by the end of April. That is it for Mr. Holland and Meteor Crater, other than one inquiry for information about the crater sent to him years later.

What follows has little to do with Meteor Crater or the drill site on its south rim. But it seems appropriate to at least draw a sketch of the remainder of his life. Mr. Holland lived in Hollywood, California which is interchangeably called Los Angeles in the documents of his Meteor Crater files. His wife and mother-in-law lived with him. During three decades they lived in two different rented homes on La Brea Avenue. That is if the census records are to be believed. The rent for the residence at 1718 La Brea was $50 per month in 1920 the period he was at Meteor Crater. By 1930 the rent there had been raised to $60 per month. In 1940 he is shown living down the street at 1768 La Brea Avenue with the rent again at $50 per month. He worked only 30 weeks and had an income of just $1300 for the entire year of 1939 as recorded in the census. The $10 per month increase in rent was two months of housing cost over a year's time. So it is likely true that they moved down the block.

Mr. Holland was well educated with five years of college. He is listed in The Mining Journal repeatedly over the years with announcements about where he is currently working and when he returns to Hollywood. He is a regular contributor to the professional publications of the mining industry. While at Meteor Crater he is contacted by T. A. Richard and asked to submit an article on the crater to Mining and Scientific Press. He lets D. M. Barringer know of the request and suggests that Barringer or perhaps Professor Magie write the article instead stating that it should not be written by a "Smart Aleck who happens to blow in here for an hour or two. . ." Quite a respectful thing for him to do since D. M. Barringer was not his boss. But he was the owner of the property where Holland worked, and a conflict of interest would likely have arisen if Holland had presented a scientific article. Mr. Jennings in Boston had even prevented Barringer's son from releasing a paper written about the crater during the years of Crater Mining Company's lease.

Mr. Holland goes out of the country to Mexico several times during his career. Once before his time at the crater in 1918. His return from that trip was at Eagle Pass, Texas. The other two returns were March 17, 1922, at Nogales, Arizona ten and a half months after his resignation at Meteor Crater. Much later on January 13, 1928, he returned from Mexico at the Naco, Arizona crossing. There was a large mining industry emerging in Mexico during those years with many American companies controlling the claims. D. M. Barringer had interests especially in iron over the years in Mexico. But for all, we know they could have been pleasure trips south of the border.

Before working at Meteor Crater, Holland had traveled and worked extensively all over the western US and also in Nova Scotia. The Evangeline Gold Mine was in full operation in 1900 soon after the company was incorporated. But the mines in the area of Halifax went bankrupt often. The gold occurs in small pockets scattered in quartz veins. It is difficult even today to mine gold in Nova Scotia profitably. In 1907 Holland was responsible for reopening the Evangeline Gold Mine.

The Hollands had a telephone at their home in Hollywood. Though it is just a lost fact of the last century, the telephone number was 577947. The census records from years long passed are always fascinating. The census takers asked the citizens some questions that would either be inappropriate or illegal today and other questions that are now wonderful glimpses into the culture of the time. The census taker in 1930 asked if there was a radio in the home. The Hollands had a radio by then. They remained in the home in Hollywood into the 1940s, but at some point, he began spending much or all of his time in Northern California working with numerous mining claims in the area around Placerville, California in El Dorado County. They were all gold claims. Some were for gold in quartz veins, and others were placer claims for sand and gravel deposits that contained gold. Most of the claims had been worked in the 1880 to 1900 period, but many saw a reopening later. California was producing approximately 1.4 million ounces of gold per year at the end of the 1930's. The area around Placerville had been responsible for about $25 million over the time it was active.

The Mining Journal reported in November of 1930 that under the supervision of L. F. S. Holland of Placerville, California, 20 men were working the property of the Guilford Gold Mining Company, on the South Fork of the American River. Two tunnels were being reopened, and a crosscut was being driven on the vein from the New York Shaft, which was 180 feet deep.

On July 24, 1939, the Nevada State Journal ran the following short notice on page four. "Preliminary work is being carried on at the Epley Mine near Placerville by a small crew of men preparatory to reopening the mine. The property comprises four patented claims and recently was taken under lease and option to purchase by the Twiford Corporation. L. F. S. Holland, mine manager, reported that the 100-foot, two-compartment shaft which meets a tunnel developed a number of years ago, will be deepened during the exploratory work. The shaft is equipped with a headframe, hoist and all necessary equipment. The quartz vein, approximately eight feet wide, lies between Mariposa slate and is described as a typical Mother Lode formation."

There was an interesting note in The Mining Journal in 1941. "A 50-ton mill will be erected at the property of the Oregon Hill Mining Company, Inc., Placerville, California. The flowsheet will include a grizzly, crusher, jig, classifier, flotation cells, and a Wilfley concentrating table. A crew of 18 men is engaged in development work in the 360-foot shaft and a tunnel near the shaft, and in preparing a site for construction of the proposed mill. The shaft has been retimbered to the 320-foot level, where a station has been cut. Drifting and crosscutting are continuing, and future plans include further development and exploration work on the 320-foot level. Reginald Owen, screen star, is president of the Oregon Hill Company, and L. F. S. Holland. Box 191, Placerville, is secretary and manager." Reginald Owens was a major actor for MGM and other studios including Disney. He had a long and distinguished acting career. The British actor moved to Hollywood in 1920 from New York and of course Mr. Holland a fellow countryman was living in Hollywood then as well.

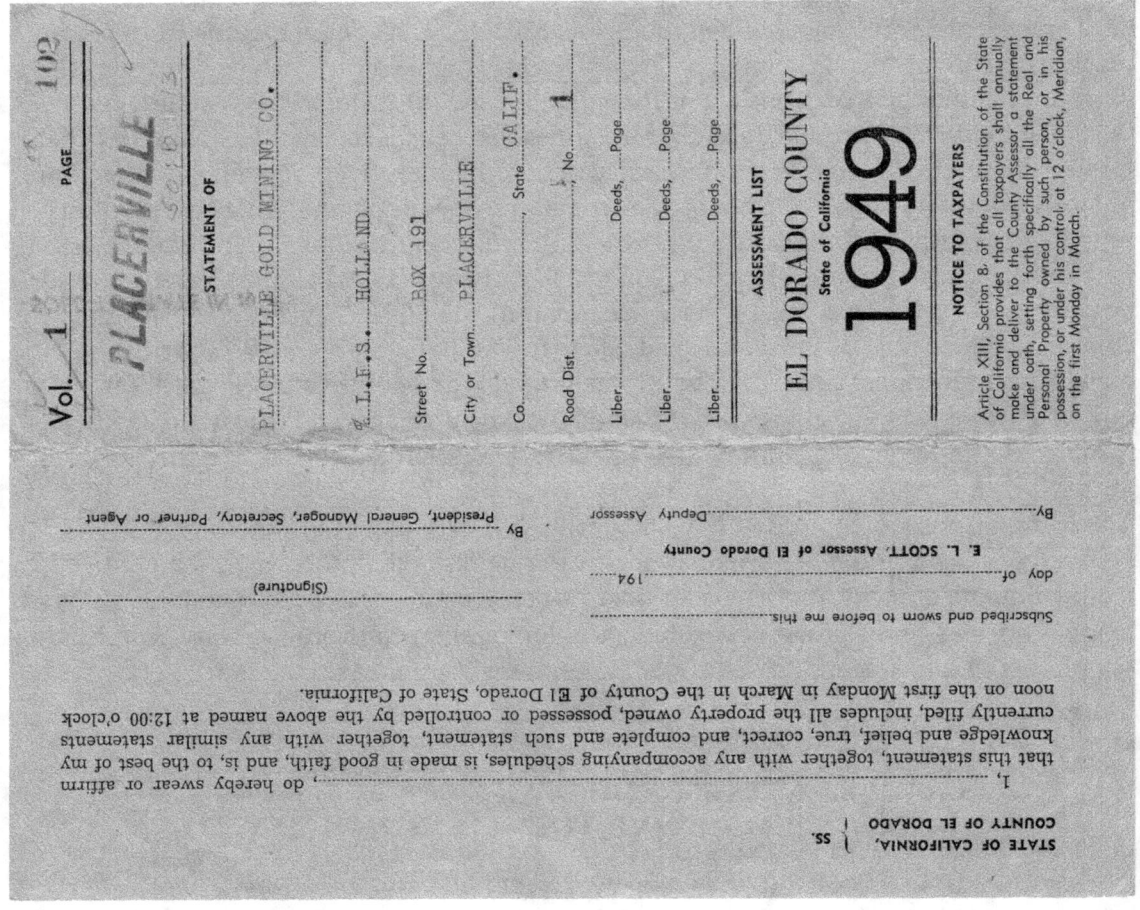

By around 1950 L. F. S. Holland has his name as the contact person for a least ten mines in the Placerville area. However, all the mines of Placerville Gold Mining Co. were likely being handled by him.

Mr. Holland never seems to have become a naturalized American citizen and is listed as an alien on all the census records. The date of his immigration varies from 1895 to 1905. He made a life claim to Social Security on the first of October 1948. It is unknown for whom the claim was made but likely it was his mother-in-law. At that time his address seems to be exclusively Post Office Box 191 in Placerville. He died in El Dorado County, California on November 13, 1957, and was cremated. His wife Lillian died in 1962 and was also cremated. They are in small niches side by side at the East Lawn Memorial Park in Sacramento, California.

Other than the presentation of his manuscript "Drilling for Meteorites" to a group of oil industry workers sometime between 1926 and 1928 there is never another mention anywhere about the time he spent at Meteor Crater. It was nearly a year in a career that lasted fifty years. Did he think much about his work there during the rest of his life? We will never know, but we can be grateful that he kept good records and that the file folders with his documents have survived for nine decades to tell us the dramatic story they contain.

Appendix One
The Drill Log of L. F. S. Holland

Log of Drill Hole Number One

From Cellar Floor to Derrick Floor distance is 18' all in Red Sandstone (Moencopie)

18' Nov. 1, 1920. Started drilling in limestone (Aubrey).
97' Nov. 3. Limestone has crevices which give tool (10 ¼") tendency to get out of line.
Nov. 4. Getting hole in shape to take 10" casing; redrilled several times, during ten days.
Nov. 14. Drilling on beam. Limestone.
109' Nov. 15. Upcast draught strong. Limestone.
115' Nov. 16. Limestone.
150' Nov. 17. Still limestone.
175' Nov. 18. Limestone.
195' to 200' Nov. 19. Entered white loose sandstone which would not hold water. Studs main frame bearing of engine broke.
Nov. 20-24. Awaiting arrival repair parts, made improvements in rig.
Nov. 25. After engine repairs made, found that during shut down of power 10" casing had become cemented tightly and that 8 ¼" casing would not go down. Hole crooked and caved.
Nov. 26-27. Trying to loosen 10" casing with black powder.
Nov. 28. Shot off shoe and 2/3 length of bottom piece of casing. Final charge 23 sticks of dynamite. Pulled free casing.
Nov. 29. Drilled as far as casing left in hole.
Nov. 30. Drilling up casing and straightening hole.
Dec. 1-6. Drilling up casing and straightening hole.
Dec. 6 Jars broke and tools lost, night tour.
Dec. 7. Made fish hook and recovered tools.
Dec. 8. Putting in 10" casing.
Dec. 9. Redrilled with jars belonging to fishing tools.
Dec. 10. Redrilling and cleaning out hole.
Dec. 11. Redrilling and cleaning out hole. Pulled 8 ¼ casing. Underreaming.
Dec. 12. New jars arrived and hitched on.
200' Dec. 13. 8 ¼ casing went to bottom.
215' Dec. 14. Limestone from 200'. Sand therefore only five feet thick.
Dec. 15. Engine down. Crank pin box could not be repaired here. Kept drilling.
250' Dec. 16. Limestone.
Dec. 17. Broken stem, jumping at square next to box and losing underreamer. Fishing.

Log of Hole Number One. Page 2

From derrick floor
250'		Dec. 18. Fishing and drilling by. Impression block shows two dents on extreme edge.
		Dec. 19. Trying to get casing over underreamer.
		Dec. 20. Drilling by.
		Dec. 21. Drilling by and fishing. Hard.
		Dec. 22. In hard limestone past the underreamer.
256'		Dec. 23. Limestone. Impression block shows obstruction half way across hole. Casing resting on underreamer.
		Dec. 24. New bit battered in two hours, another bit broke.
257'		Dec. 25. Stem broke, upper end at square. Fished and recovered in three hours. Very hard limestone.
260'		Dec. 26. Considerable chips of iron in bailer, from lost underreamer probably. Limestone.
		Dec. 27. Third stem broke at same place as others, at square next to box. Lost bit.
264'		Dec. 28. Drilling by.
266'		Dec. 29. Limestone in bailer.
		Dec. 30. Ten foot sinker, all there was left to drill with, broke at square at top end. Removed bit and stem 6 p.m.
		Dec. 31. Going over engine.

1921
		Jan. 1. Dressing bits and repairs.
		Jan. 2. Drilling without stem.
		Jan. 3. Drilling without stem. Impression block shows two dents.
270'		Jan. 4. Impression block showed curve. Shot twice with dynamite. Bit apparently standing on underreamer.
		Jan. 5. Drilling by and fishing.
		Jan. 6. Drilling by and fishing.
		Jan. 7. Drilling by and fishing.
272-273'	Jan. 8. Drilling by and fishing. Impression block showed dent on edge only. Limestone and iron.
275'		Jan. 9. Drilling by and fishing. 30' repaired stem hitched on at 3 p.m. On arrival. Limestone with considerable steel and iron from tools.
276'		Jan. 10. Bailer went to 276' 6". Limestone with steel from tools. Impression block showed lug of underreamer, apparently upside down. Bit may have been drilled by. Bit came unscrewed. Lost.
277'		Jan.11. Fishing with three pronged grab etc. Drilling by.
		Jan. 12. 262' 6" of 8 ¼" casing pulled. Shoe found to be broken off. Bailer would not go beyond 200'. After cleaning out hole, large impression block showed shoe only. Fished for it.

Log of Hole Number One Page 3

From derrick floor
277'	Jan. 13. Impression block showed shoe and mass in middle of irregular shape. Fished No results. Hitched on 6 5/8" bit for drilling. Went down very slowly.
282'	Jan. 14. Bailer to 278' only. Progress only a few inches with 8 ¼" bit. At night string of tools got stuck and was released by drilling alongside with 6 5/8" bit on sand line.
	Jan. 15. Plenty of steel and limestone in Bailer, also some sand which may have fallen in from above. Jars broke losing jars and bit.
	Jan. 16. Bit hook ordered in December arrived and hitched on.
	Jan. 17. Bit lost on 10th recovered. Afternoon recovered bit and jars lost on 15th.
	Jan. 18. Fishing without results. 30' stem pin bound and broken at box.
	Jan. 19. Fished hole for 30' and redrilled with 10" bit. Four bailers of water made on 4' in hole.
	Jan. 20. Drilling by.
284'	Jan. 21. Drilling by.
286'	Jan. 22. Drilling by. Limestone. 8 ¼" casing put in, reaching 287'. Very little progress made with drilling in night, having no stem.
	Jan. 23. Drilling up some iron or steel in bottom. Progress slow with no stem.
288' Jan. 24. Bailings showed some white sand. Impression block showed flat piece half area of block. Steel in bailings.
289' Nearly pure sandstone in bailings. Drilling slowly with no stem.
  Jan. 25. Impression in block looks like that of box off tool, apparently jumping around on its side in the hole.
290'	Unwashed bailing shows lime and sand with considerable iron probably from tools. Iron shows no reaction for nickel. Fishing tools fixed up in effort to pick up box without success. Left casing on bottom of hole and drilled.
	Jan. 26. Something fell into hole, perhaps underreamer. Pulled casing and fished.
	Jan. 27. Bit hook caught in 10" casing which was pulled. Bailer to 291'.
291'
293'	Jan. 28. Drilling by and fishing. Pieces of coarse white sandstone but mostly brown and yellow stained sandstone. No reaction for nickel.
295'	Jan. 29. Sandstone. Drilling by.
297'	Jan, 30. Sandstone. Drilling impeded by loose box in hole.
300'	Jan. 31. Sandstone. *(unreadable words going off bottom of page)*

Log of Hole Number One. Page 4.

From derrick floor

300'   Feb. 1. The piece of iron moving around in the hole continued to interfere with drilling. It could not be fished out.

307'   Feb. 2. Took off manila rope and hitched 30' stem on wire cable. Put back casing.

Feb. 3. Reached bottom of hole at 8 p.m.

310'   Feb. 4. Fished out broken 8 ¼" bit. Thirteen inches off.

312'   Feb. 5. Spud arrived and hitched on. Impression block showed very little of the piece broken off bit, but the latter is moving around in the hole impeding drilling.

Feb. 6. Tried to drill by, and spudded.

Feb. 7. Trying to spud by and fishing.

Feb. 8. Trying to spud by and fishing.

Feb. 9. Foreman Wammock decided that there was no chance of drilling by or getting spud by, his belief being that the underreamer was broken in two and was lying on the bottom of the hole, completely blocking it. Instructions ultimately received from Sidney J. Jennings were to drive a tunnel from the Crater at least ten feet below the bottom of the drill hole to get out the lost tools from it.

Superintendent.

APPENDIX TWO

Drill Log Crater Mining Company Hole Number One
C. W. Plumb Superintendent

Records from USSR&M Co.

Summary of work by Superintendent L. F. S. Holland
18 feet Bottom of Cellar
174 Limestone. Many crevices in limestone
174-195 {Interval not logged}
195-200 Sandstone. Lighted center goes out of sight. White, loose
200-250 Limestone. Drilling past broken cores
250-257 Limestone. Particles of shale ball
257-283 Limestone
282-288 Limestone. Sand commenced to show in bailings
288-290 Brown and yellow stained saccharoidal sand. Much iron and steel, no nickel
290-311 Sand
311-312 Lost hole. Underreamer at bottom

C. W. Plumb Superintendent commences restart of drilling

312-326 In tunnel 71–vesicular sandstone
326-380 White sandstone, very quick. 375' hard material dropped in hole
380-384 Drill twisted off
390-425 White sandstone
425-460 White sandstone
460-464 Red cong. Or coarse red sandstone. Small shells, 1/8" long
464-468 Red clay–streaks of white calcite
468-480 Red or cong. or coarse red sandstone
480-500 Red sandstone or conglomerate. Steel
500-520 White sandstone
600-603 White sandstone
603-605 Hard conglomerate
605-607 Hard conglomerate
607-615 Possibly white sandstone with reddish streaks
615-620 Hard conglomerate
620-627 Whitish sandstone
627-669 White sandstone
669-684 White sandstone (?), red sandstone at bottom
684-725 Red sandstone. Hard to get samples
725-750 Crevice. Coarse grains of limestone, sandstone, and silica
750-801 Red mud or clay, white streaks
801-820 Red clay with calcite streaks

820-876 Large crevice. Rounded pebbles of limestone, silica, and sandstone. Iron nodules, all cemented
876-930 Same conglomeratic material
930-940 Conglomerate as above
940-942 Red clay
942-953 Red clay
953-957 Iron nodule. No nickel
957-1096 Alternating layers of white and gray sandstone. Drilled easily
1096-1100 Hard drilling, iron nodule, no nickel
1100-1130 Hard nodules, similar to 958. Segregations from sandstone, filled with small black particles–silicon, effervesces readily
1130-1287 Hard boulders in siliceous white sandstone. 1" to 6". Some nodules show nickel reaction–perhaps shale balls. Some greenish material, looks like clay
1130-1134 4' very hard, like rest of boulders
1134-1145 Soft sandstone. Small greenish pieces of metal or slag in sample. Slight show of nickel
1145-1168 Hard and soft material, slight nickel, layers 6 inches to one-foot layers
1168-1187 Very soft, white silica sand. Then hard and white like silica sand found in crater
1187-1188 Five hours, sample very black, heavy, greenish pieces of metal, very strong nickel
1188-1190 Same as above
1190-1208 First foot hard. Then alternate hard and soft in 6 inches to one foot
1208-1228 Very soft for 15', then hard and rough. Good test of nickel. Silica sand almost transparent
1228-1251 2' hard, rough. Five hours on last foot, stray nickel
1231-1235 4' hard, rough
1235-1249 Silica sand, medium soft. Slight nickel
1249-1271 Soft , white sandstone. Hard nodules at 1,255', 1,260', 1,270'. No nickel
1271-1276 Hard and rough. Like nest of hard boulders. Fine nickel test
1276-1287 Easier for 5 feet. Then harder and rough, fine nickel test
1287-1293 Drilled very hard 4', hard to get samples. Then easier, good nickel
1293-1311 Hard few inches. Then very soft. Fair nickel test
1311-1323 Easy drilling 10', then very rough. Samples quite black. White sandstone and black material about 50 per cent each. Few pieces red sandstone
showing. Shells. White sandstone getting harder. Samples show good nickel test
1323-1335 Drilling rough for 7 feet. Then smooth and very hard. Many pieces of hard red sanstone. Also many shells 1/8" long. Fine nickel test
1335-1339 Reamed very hard, like in boulders size of baseball. Drillings looked very black. Samples all gave fine nickel test, about 75 per cent mineral
1339-1350 Drilling hard but smooth. Some red sandstone but mostly black or brownish pieces of material, very magnetic. Best nickel test yet
1350-1352 Hard for 2 feet. Lost sludge at once. Lost circulation. From 1,095' to 1,352' black mineral particles, plentiful

1352-1360 Formation about as last 250 feet. Nickel about same
1360-1370 Formation hard and rough. Shale ball appearance. Last 25 feet
1370-1376 Extremely hard and rough. Strong nickel test. Samples look as if we are passing through a recemented mass of conglomerate as we find shells, rounded pebbles of red sandstone and of limestone, and also a great many small brown pieces resembling shale balls. Stuck and had to abandon at
1,376 feet. Bit appears to have wedged under boulders

End of Log

Some Final Words on Appendix Two

This log is different from Mr. Holland's. It has no dates for the entries. It is simply a record of the type of material drilled through at the various depths. An entry occurs at intervals or when the material changes, or if something of interest is found in the cuttings from the hole. Sampling at deeper levels includes increasingly frequent tests for nickel.

The Holland portion of the log is simplified by Plumb in a sort of "its old news" manner. Only a few entries are made for the first 312 feet. The first 195 feet of drilling has been compressed into just three entries, but there is an insight into how the work was done that was not clearly described by Holland. The lantern the drillers sent down the pipe which disappeared into the cavernous fissure is a fascinating story to read. However, the lantern was likely put down the hole to determine if the hole was straight not to check for such a fissure. If the light which was mechanically made to stay in the center of the hole disappeared it could be seen going out of line to one side from the surface. This would be an indication in which direction the hole was going crooked. Plumb notes this happening at the 195-200 foot depth, but this is not in Holland's log entry for that date. Holland reports "Entered white loose sandstone which would not hold water. Studs main frame bearing of engine broke." Plumb had some resources for notes besides just a copy of Holland's published log. Drillers often kept logs and notes as well at the rig. A verbal "turn over" from shift to shift may have been done but also written notes on a log sheet were probably made as well. Holland may have been collating that material and picking the items to report weekly to Boston while other details were omitted. Plumb has offered a nice bit of instruction for us today about how crooked holes were noticed a hundred years ago.

Little mention is made about difficulties in the actual drilling other than the notes as to the hardness or roughness of the rock and the cutting. The period of this log from 326 feet to the end at 1,376 feet covers a timespan of about 17 months. It is, therefore, reasonable to conclude as has been written in other books that Plumb had great difficulties in the work just as Holland experienced. Only the one mention of the bit twisting off is listed. The often small progress made in the entries and the notes of the hardness of the material further support that this was nearly as difficult a drilling operation after Holland was replaced. The criticism by Barringer certainly does not exclude Mr. Plumb and is, in fact, more directed at him. He was in charge when the drill reached the more important depths and when it failed just short of the goal of undisturbed rock leaving no clear answers for Barringer.

From 1187 feet in depth to the end of the log the entries are even more interesting. For a long distance, the material alternates from very hard to very soft. There are many entries showing good nickel tests. It has been theorized that the broken rocks of the region surrounding the initially formed crater were injected with melted asteroid material. This might explain the rapidly alternating layers of hard and soft rocks the bit encountered. Testing showed they were rich in nickel and then free of nickel.

There is one mention of steel in the hole and of a hard material falling into the hole. These were early entries, and it is not beyond possibility that there was still some of the Holland phase steel and iron from the hole above making its way to the lower level. However, the big opening underground at the end of the tunnel would seem to make this a lower possibility. Unless it was occurring somehow during times that the casing was being pulled. Then some of Holland's sidetracked and bypassed tools, shoe pieces, and blasted casing pieces might again be free to cave off and fall into the deeper level. The steel may have been from problems Plumb experience not recorded in this log. Furthermore, they would have redrilled the hole completely with the new rotary bit and thereby cleaned the hole of most of any remaining debris from Holland's work.

Whether or not the end of the tunnel was used as an access point for changing tools of other activities has never been mentioned. The numbers do not seem to be big enough to allow for such work underground. The tunnel end was only opened up 14 feet in height not enough to handle churn drill stems, for rotary equipment the answer is an unknown. But the tunnel mouth is not near the derrick at all. Instead, it is a long hike down into the crater. It is unlikely that the tunnel served a further purpose after the broken underreamer was removed.

There is no doubt about continuing fissures and cavities in the rock they are mentioned in entries in the Plumb log. One of the most interesting entries is quoted here "1350-1352 Hard for 2 feet. Lost sludge at once. Lost circulation. From 1,095' to 1,352' black mineral particles, plentiful" It did not end in a period, so none has been included in the quote. It must be read however observing the rest of the punctuation. There was a loss of the water supply. No sludge or mud as Holland had called it was coming to the surface. In rotary drilling, the water might have been pumped through the drilling equipment to flow out the top of the hole. The next phrase "Lost circulation" makes it clear that the water has stopped moving. But then there is a period and the next word begins a new sentence. "From 1,095' to 1,352' black mineral particles, plentiful" cannot be connected to the previous phrase to indicate that the hole remained dry for that distance. Nothing more than that the circulation stopped is stated. It is conceivable that the workers found the difficulty with the pump and fixed it and continued drilling normally. If this is the source of the idea that the hole was dry for 57 feet near the bottom which Barringer mentions as part of his belief that it is safe to dig on the south slope, it may be the source of a mistaken idea. However, as stated before the weekly reports of Plumb may clearly indicate that it was indeed a dry hole at some point or that the loss of circulation was the result of the hole draining and not the fault of equipment. But that would have been a very significant problem and it would have made the sampling and testing impossible since no material would have come to the surface without water flowing. His next statement in the same entry is about black mineral particles being plentiful. Then later at 1350 feet Plumb notes the "best nickel tests yet" were found in the cuttings coming to the surface. It is hard to imagine that if the water circulation problem had continued that further mention would not have been made by Plumb in the log. It should have been the highlight remark for the time it was happening or should have stopped the drilling altogether. Since there are no dates in his log, it is possible it did stop the drilling until a repair was made.

This author has read the last entry of the Plumb log many times in thirty years. The continuing mention of small shells and red sandstone in the material still seems to indicate that there was never a large mass of asteroid hit by the bit, just material filling cracks and maybe as recently suggested melted iron injected into cracks which is now weathered iron shale like mineral. Just the two spots where the sludge material is green and metal is noted. Levels which took five hours or more to drill a single foot still seem to be meteorites that were hit. Each location was only several feet thick.

This is all beyond the scope of a book about Mr. Holland and his time at Meteor Crater. It is connected though, and worthy of mention for it closes up the work which Holland was not able to complete himself.

www.ingramcontent.com/pod-product-compliance
Lightning Source LLC
Chambersburg PA
CBHW082325220526
45470CB00008B/2399